I0878372

THE POSTDOCTORAL EXPERIENCE REVISITED

Committee to Review the State of
Postdoctoral Experience in Scientists and Engineers

Committee on Science, Engineering, and Public Policy

Policy and Global Affairs

NATIONAL ACADEMY OF SCIENCES,
NATIONAL ACADEMY OF ENGINEERING, *AND*
INSTITUTE OF MEDICINE
OF THE NATIONAL ACADEMIES

THE NATIONAL ACADEMIES PRESS
Washington, D.C.
www.nap.edu

THE NATIONAL ACADEMIES PRESS 500 Fifth Street, NW Washington, DC 20001

NOTICE: The project that is the subject of this report was approved by the Governing Board of the National Research Council, whose members are drawn from the councils of the National Academy of Sciences, the National Academy of Engineering, and the Institute of Medicine. The members of the committee responsible for the report were chosen for their special competences and with regard for appropriate balance.

This study was supported in part by Award Number 2012-10-22 between the National Academy of Sciences and the Alfred P. Sloan Foundation. Any opinions, findings, conclusions, or recommendations expressed in this publication are those of the authors and do not necessarily reflect the views of the organizations or agencies that provided support for the project.

International Standard Book Number 13: 978-0-309-31446-6
International Standard Book Number 10: 0-309-31446-1
Library of Congress Control Number: 2014957852

Additional copies of this report are available from the National Academies Press, 500 Fifth Street, NW, Room 360, Washington, DC 20001; (800) 624-6242 or (202) 334-3313; http://www.nap.edu

Printed in the United States of America

THE NATIONAL ACADEMIES

Advisers to the Nation on Science, Engineering, and Medicine

The **National Academy of Sciences** is a private, nonprofit, self-perpetuating society of distinguished scholars engaged in scientific and engineering research, dedicated to the furtherance of science and technology and to their use for the general welfare. Upon the authority of the charter granted to it by the Congress in 1863, the Academy has a mandate that requires it to advise the federal government on scientific and technical matters. Dr. Ralph J. Cicerone is president of the National Academy of Sciences.

The **National Academy of Engineering** was established in 1964, under the charter of the National Academy of Sciences, as a parallel organization of outstanding engineers. It is autonomous in its administration and in the selection of its members, sharing with the National Academy of Sciences the responsibility for advising the federal government. The National Academy of Engineering also sponsors engineering programs aimed at meeting national needs, encourages education and research, and recognizes the superior achievements of engineers. Dr. C. D. Mote, Jr., is president of the National Academy of Engineering.

The **Institute of Medicine** was established in 1970 by the National Academy of Sciences to secure the services of eminent members of appropriate professions in the examination of policy matters pertaining to the health of the public. The Institute acts under the responsibility given to the National Academy of Sciences by its congressional charter to be an adviser to the federal government and, upon its own initiative, to identify issues of medical care, research, and education. Dr. Victor J. Dzau is president of the Institute of Medicine.

The **National Research Council** was organized by the National Academy of Sciences in 1916 to associate the broad community of science and technology with the Academy's purposes of furthering knowledge and advising the federal government. Functioning in accordance with general policies determined by the Academy, the Council has become the principal operating agency of both the National Academy of Sciences and the National Academy of Engineering in providing services to the government, the public, and the scientific and engineering communities. The Council is administered jointly by both Academies and the Institute of Medicine. Dr. Ralph J. Cicerone and Dr. C. D. Mote, Jr., are chair and vice chair, respectively, of the National Research Council.

www.national-academies.org

COMMITTEE TO REVIEW THE STATE OF POSTDOCTORAL EXPERIENCE IN SCIENTISTS AND ENGINEERS

GREGORY A. PETSKO (*Chair*), Arthur J. Mahon Professor of Neurology and Neuroscience, Weill Cornell Medical College; Gyula and Katica Tauber Professor of Biochemistry and Chemistry Emeritus, Brandeis University
SIBBY ANDERSON-THOMPKINS, Director of Postdoctoral Affairs, University of North Carolina, Chapel Hill
H. RUSSELL BERNARD, Professor Emeritus, Department of Anthropology, University of Florida
CAROL GREIDER, Daniel Nathans Professor and Director, Department of Molecular Biology and Genetics, Johns Hopkins University School of Medicine
JAMES PLUMMER, Frederick Emmons Terman Dean of the School of Engineering, and John M. Fluke Professor of Electrical Engineering, Stanford University
E. ALBERT REECE, Vice President for Medical Affairs, Bowers Distinguished Professor and Dean of the School of Medicine, University of Maryland, Baltimore
NANCY SCHWARTZ, Professor of Pediatrics and Biochemistry, Dean for Postdoctoral Affairs, The University of Chicago
PAULA STEPHAN, Professor of Economics, Georgia State University
LORRAINE TRACEY, Medical Science Liaison, Teva Pharmaceuticals
MICHAEL TURNER, Rauner Distinguished Service Professor and Director, Kavli Institute for Cosmological Physics, The University of Chicago
ALLISON WOODALL, Deputy General Counsel, Labor, Employment and Benefits Group, University of California System
JOAN WOODARD, Retired Executive Vice President and Deputy Director, Sandia National Laboratories

Staff
KEVIN FINNERAN, Study Director
GURUPRASAD MADHAVAN, Senior Program Officer
MARIA LUND DAHLBERG, Associate Program Officer
JAMIE BIGLOW, Senior Program Assistant
MARION RAMSEY, Administrative Associate (until October 2013)
DAVID PROCTOR, Research Associate (until September 2012)

Consultant Writers
DAVID PROCTOR
JOHN CECCATTI

COMMITTEE ON SCIENCE, ENGINEERING, AND PUBLIC POLICY

RICHARD N. ZARE [NAS] (*chair*), Marguerite Blake Wilbur Professor, Stanford University
LINDA M. ABRIOLA [NAE], Dean of Engineering, Tufts University
SUSAN ATHEY [NAS], Professor, Graduate School of Business, Stanford University
MOSES H. W. CHAN [NAS], Evan Pugh Professor of Physics, Pennsylvania State University
RALPH J. CICERONE [NAS] (*ex-officio*), President, National Academy of Sciences
PAUL CITRON [NAE], Vice President (retired), Technology Policy and Academic Relations, Medtronic, Inc.
DAVID DANIEL [NAE], President, The University of Texas at Dallas
GORDON R. ENGLAND [NAE], President, E6 Partners LLC
VICTOR J. DZAU [IOM] (*ex-officio*), President, Institute of Medicine
DIANE E. GRIFFIN [NAS, IOM], Alfred and Jill Sommer Professor, Chair in Molecular Microbiology and Immunology, Johns Hopkins Bloomberg School of Public Health
JOHN G. HILDEBRAND [NAS], Regents Professor, Department of Neuroscience, University of Arizona
DAVID KORN [IOM], Professor of Pathology, Massachusetts General Hospital and Harvard Medical School
C. D. MOTE, JR. [NAE] (*ex-officio*), President, National Academy of Engineering
PERCY A. PIERRE [NAE], Vice President and Professor Emeritus, Michigan State University
E. ALBERT REECE [IOM], Vice President for Medical Affairs, Bowers Distinguished Professor and Dean, School of Medicine, University of Maryland, Baltimore
MICHAEL S. TURNER [NAS], Rauner Distinguished Service Professor, Kavli Institute for Cosmological Physics, The University of Chicago
NANCY S. WEXLER [IOM], Higgins Professor of Neuropsychology, Colleges of Physicians and Surgeons, Columbia University
PETER WOLYNES [NAS], D.R. Bullard-Welch Foundation Professor of Chemistry, Center for Theoretical Biological Physics-BCR, Rice University

Staff
KEVIN FINNERAN, Director
TOM ARRISON, Senior Program Officer
GURU MADHAVAN, Senior Program Officer
MARIA LUND DAHLBERG, Associate Program Officer

JAMIE BIGLOW, Senior Program Assistant
NEERAJ GORKHALY, Research Associate (until February 2014)
MARION RAMSEY, Administrative Associate (until October 2013)
DAVID PROCTOR, Research Associate (until September 2012)

Preface

In 2000, the National Academies' Committee on Science, Engineering, and Public Policy (COSEPUP) released a report, *Enhancing the Postdoctoral Experience for Scientists and Engineers*, that examined the experience of postdoctoral researchers in the United States. The report stressed that a rapid expansion of postdoctoral training during the previous decade had taken place without adequate oversight, resulting in fundamental changes in the nature of the experience for many postdoctoral researchers. It offered guidelines for improving the postdoctoral system, with specific directions for a range of stakeholders: federal agencies, universities, foundations, professional organizations, and postdoctoral researchers themselves. In the almost 15 years since the release of this report, the number of postdoctoral researchers in all research disciplines continued to grow sharply, whereas the number of independent and especially academic research positions into which they might transition did not. The National Academy of Sciences, the Institute of Medicine, and the National Academy of Engineering felt that, in the light of these developments, another examination of the postdoctoral experience was necessary and timely. In particular, it seemed desirable to examine the available data on the number of postdoctoral researchers in various disciplines and types of institution, their national origins and means of support, their salaries, degree of satisfaction, and their career outcomes, and to determine the extent to which the recommendations of the 2000 COSEPUP report had been implemented and their effects on the overall experience.

The present report is an attempt at such an examination. It compiles and analyzes the best publicly available data and considers how the recommendations of the earlier report have affected the behavior of institutions and individual postdoctoral researchers. It uses these considerations to recommend further steps that all the participants in the research enterprise can take to improve the quality of postdoctoral experiences, and lays out a set of best practices toward achieving these recommendations. In formulating these guidelines, the present committee was guided by one general principle: that the postdoctoral period should be a defined period of advanced training and mentoring in research, and that it should also be, as the majority of the committee members remembered from their own experience, among the most enjoyable times of the postdoctoral researcher's professional life.

Acknowledgments

This report has been reviewed in draft form by individuals chosen for their diverse perspectives and technical expertise, in accordance with procedures approved by the National Academies' Report Review Committee. The purpose of this independent review is to provide candid and critical comments that will assist the institution in making its published report as sound as possible and to ensure that the report meets institutional standards for objectivity, evidence, and responsiveness to the study charge. The review comments and draft manuscript remain confidential to protect the integrity of the process.

We wish to thank the following individuals for their review of this report: David Allison, University of Alabama, Birmingham; Roger Cone, Vanderbilt University; Geoff Davis, Google, Inc.; John Dowling, Harvard University; Leroy Fletcher, Texas A&M University; Keith Micoli, New York University Langone Medical Center; Georgine Pion, Vanderbilt University; Samuel Preston, University of Pennsylvania; Mary Thomas, Bill and Melinda Gates Foundation; Shirley Tilghman, Princeton University; Gerlind Wallon, European Molecular Biology Organization; and Ward Winer, Georgia Institute of Technology.

Although the reviewers listed above have provided many constructive comments and suggestions, they were not asked to endorse the conclusions or recommendations, nor did they see the final draft of the report before its release. The review of this report was overseen by George Langford, Syracuse University and Julia Philips, Sandia National Laboratories. Appointed by the National Academies, they were responsible for making certain that an independent examination of this report was carried out in accordance with institutional procedures and that all review comments were carefully considered. Responsibility for the final content of this report rests entirely with the authoring committee and the institution.

Gregory A. Petsko
Chair

Contents

Summary

Concern about the postdoctoral training system has been gnawing at the research community for decades. The National Academies produced reports in 1969, 1980, and 2000 that called for reforms to the system. In the past 5 years, the National Institutes of Health (NIH), the President's Council of Advisors on Science and Technology, the American Chemical Society, the Council of Graduate Schools, among others have expressed their concerns in reports.

The sources of uneasiness have changed only slightly over time. Is it really necessary for someone to remain in training until their mid-30s before being qualified for his or her chosen career track? Are these highly qualified Ph.D. researchers receiving the recognition and remuneration that they deserve? Is there an appropriate balance between the number of postdoctoral researchers that are trained and the number of jobs that require postdoctoral training? What happens to those postdoctoral researchers who do not get the jobs they aspire to? How concerned should the research enterprise be with the postdoctoral researchers in the United States on temporary visas? Who is responsible for ensuring that postdoctoral researchers are treated fairly and receive the mentoring and training that is essential to their position? Is the perceived status of postdoctoral researchers a possible disincentive to undergraduates and graduate students who are considering independent research careers? Why are more reliable data not collected about the current postdoctoral population and the career outcomes of former postdoctoral researchers?

What has changed is the percentage of Ph.D.'s who pursue postdoctoral training. It is growing steadily and spreading from the biomedical and physical sciences to engineering and the social sciences. Although the data are not definitive, the average length of time spent in postdoctoral positions seems to be increasing. The sources of funding have also changed. The number of postdoctoral fellowships and traineeships, which provide postdoctoral researchers relative autonomy and recognition, has remained nearly constant for decades, whereas the number of postdoctoral researchers hired as part of research grants or supported by non federal sources has grown dramatically.

For some postdoctoral researchers the system works very well. They gain valuable research experience and career guidance from an accomplished researcher. They learn to develop ideas for independent research, apply for grants, and manage a lab; they cultivate professional networks and publish papers. They eventually move into tenure-track research faculty positions at leading universities. However, it is known that this is not the norm because the growth in the number of postdoctoral researchers far exceeds the growth in the number of tenure-track job openings. Unfortunately, there is not clear definition for what the norm is because, in spite of repeated calls for better tracking data, the picture of the postdoctoral experience remains foggy. Whereas aggregate trends can be discerned, only rough estimates of the total number of postdoctoral researchers, and no good information about what becomes of the postdoctoral researchers who earned their Ph.D.'s outside the United States, exist.

One important finding is that the postdoctoral experience differs considerably among types of institutions. Compared to postdoctoral researchers working at universities, postdoctoral researchers who work at national labs or in industry are typically paid much more, remain for shorter periods, and are often offered fulltime jobs at the end of their appointment. Likewise, postdoctoral researchers who are on fellowships or traineeships have higher salaries, better mentoring, and more control over their research than those who are working under a principal investigator's research grant. The majority of postdoctoral researchers are working under research grants.

The period since the National Academies' 2000 report *Enhancing the Postdoctoral Experience for Scientists and Engineers* (2000 Postdoctoral Report) has seen a number of significant advances in the treatment and understanding of postdoctoral researchers. Many universities have created offices of postdoctoral affairs to provide better services to postdoctoral researchers. The postdoctoral researchers created the National Postdoctoral Association (NPA), which includes representatives of the offices of postdoctoral affairs, to provide information to postdoctoral researchers, a forum for discussion, and a unified voice. The NIH created an office of postdoctoral affairs for its intramural postdoctoral researchers. The National Science Foundation (NSF) has added a requirement that research proposals that include hiring a postdoctoral researcher include a mentoring plan. Organizations such as Sigma Xi and the American Association of Universities conducted surveys of postdoctoral researchers. The Association of American Medical Colleges created the Graduate Research, Education, and Training (GREAT) Group to address questions about postdoctoral training. Several individual researchers have conducted studies. The American Association for the Advancement of Science developed *MyIDP*, software that enables graduate students and postdoctoral researchers to develop individual development plans that help them to understand their career options more clearly and to make better informed career decisions.

Other aspects of postdoctoral training have seen little change. The quality of data about postdoctoral researchers is still insufficient. Although postdoctoral affairs offices have made efforts to improve mentoring, there is no convincing evidence that most postdoctoral researchers are receiving adequate mentoring. In spite of the fact that a very large percentage—a majority in some fields—of postdoctoral researchers eventually pursue careers other than that of tenure-track research faculty, there is little evidence that universities and mentors are providing adequate information about and preparation for other types of careers. Salaries, which have always been relatively low, have failed to even keep pace with inflation. Somewhat surprisingly, several surveys have found that although postdoctoral researchers would prefer better pay, they are even more concerned about their lack of recognition, status, and structured programs. The running gag in the comic strip "Postdoc Funnies" is that postdoctoral researchers are invisible. A fear expressed by many of the postdoctoral researchers who testified to the committee is that they will be able to move only into even less visible positions as university staff scientists or adjunct faculty.

Although the postdoctoral researchers themselves might feel invisible, there is broad recognition that something is amiss in the postdoctoral training system. Lack of data makes it difficult for leaders in research institutions and funding agencies to make policies about the role postdoctoral training should play in the research enterprise and for young people interested in science and engineering to make informed decisions on their career paths. Most postdoctoral researchers are employed by principal investigators to work on research grants, which creates an inherent source of stress. The investigator's primary mission is to complete the research, and any time spent in training the postdoctoral researcher is time not spent on the research. Principal investigators play an essential role in the training of postdoctoral researchers that must be acknowledged and reinforced by the funding agencies and institutions. Postdoctoral researchers are the future of the research enterprise, so it is critical that this period of training attract the most capable people to research.

This report focuses on academic postdoctoral researchers because they are by far the largest component of the population, because less is known about postdoctoral researchers in industry, and because what we do know indicates that postdoctoral researchers in industry and at national laboratories do not face the same problems as academic postdoctoral researchers. Their roles are better defined, salaries are higher, terms are shorter, and the connection to career development is clearer. For this reason the recommendations that follow are intended to address the problems primarily encountered by postdoctoral researchers in the academic setting.

Using a definition of a postdoctoral researcher agreed upon by the NPA, NIH, and NSF as a guide—"An individual who has received a doctoral degree (or equivalent) and is engaged in a temporary and defined period of mentored advanced training to enhance the professional skills and research independence needed to pursue his or her chosen career path"—the committee has developed recommendations for best practices covering five aspects of the postdoctoral experience: period of service, title and role, career development, compensation and benefits, and mentoring. In addition, the committee stresses the importance of data collection through a sixth recommendation. While the recommendations are numbered, this is for ease of reference and should not be taken to imply prioritization; these six items are necessarily interconnected.

1. *Period of Service*: The committee endorses the recommended practice, put forward by the National Institutes of Health, the National Science Foundation, and the National Postdoctoral Association in 2007, that postdoctoral research training is and should be a "temporary and defined period." **Postdoctoral appointments for a given postdoctoral researcher should total no more than 5 years in duration, barring extraordinary circumstances.** This maximum term should include cumulative postdoctoral research experience, though extensions may be granted in extraordinary circumstances (e.g. family leave, illness).

This recommendation requires direct actions by the host institutions and the funding agencies.

1.1 Host institutions should maintain a record of how long a postdoctoral researcher remains in a position and provide that information to funding agencies as part of grant proposals.

1.2 To facilitate tracking of postdoctoral researchers, funding agencies could assign each postdoctoral researcher an identifier and keep a record of the total length of time any given individual is holding such a position.

2. *Title and Role*: In many instances, positions currently occupied by postdoctoral researchers are more appropriately filled by permanent staff scientists (e.g., technicians, research assistant professors, staff scientists, laboratory managers). **The title of "postdoctoral researcher" should be applied only to those people who are receiving advanced training in research.** When the appointment period is completed, the postdoctoral researchers should move on to a permanent position externally or be transitioned internally to a staff position with a different and appropriate designation and salary.

This recommendation requires action primarily by the funding agencies and the host institutions.

2.1 Funding agencies should have a consistent designation for "postdoctoral researchers," and require evidence that advanced research training is a component of the postdoctoral experience.

2.2 Host institutions should create or identity professional positions for individuals who are conducting research but who are not receiving training, and these individuals should receive appropriate remuneration, benefits, and privileges.

3. *Career Development*: **Host institutions and mentors should, beginning at the first year of graduate school, make graduate students aware of the wide variety of career paths available for**

Ph.D. recipients, and explain that postdoctoral positions are intended only for those seeking advanced research training. Career guidance should include, where feasible, the provision of internships and other practical experiences. **The postdoctoral position should not be viewed by graduate students or principal investigators as the default step after the completion of doctoral training.**

This recommendation requires action by all the different members of the research system: the funding agencies, the host institutions, the professional societies, the mentors, the postdoctoral researchers, and even the graduate students before becoming postdoctoral researchers.

3.1 Host institutions, especially those with graduate student populations, should provide multiple engagement activities to help students explore all avenues of career development. Funding agencies should help to support these efforts.

3.3 Professional societies should gather and disseminate information about the full range of career paths within their discipline. Useful activities could include collecting statistics about job openings and salaries, identifying individuals in various sectors who can provide career advice, and organizing career fairs at professional meetings.

3.3 Mentors, in addition to providing guidance based on their own experience, should become familiar with and disseminate information about all forms of career development opportunities available either at the host institution or through their professional society.

3.4 Postdoctoral researchers and graduate students have a responsibility to participate in the career development opportunities provided by their institutions, to explore other sources of information such as professional societies, and to use available career-development tools.

4. ***Compensation and Benefits of Employment***: Current postdoctoral salaries are low. Salaries should be increased to (1) reflect the qualifications of postdoctoral scholars, (2) address the slow progress the community has made toward implementing salary increases as recommended in several National Research Council reports, and (3) adjust the relative wage of postdoctoral researchers to appropriately reflect their value and contribution to research. The committee considered five different approaches for determining an appropriate minimum salary: (1) indexing to contemporary college graduates, (2) indexing to graduate stipends, (3) indexing to newly hired assistant professors, (4) inflation of previous recommendations, and (5) Research Grade Evaluation Guide. All of these approaches, which are discussed in detail in Appendix B, suggest an amount of $50,000 or

more. In addition, despite considerable variation in salaries by field, geographic area, and sector, data on starting postdoctoral salaries reveal that the starting salary prescribed by the National Institutes of Health (NIH) for the Ruth L. Kirschstein National Research Service Award (NRSA) postdoctoral award (currently set at $42,000 for 2014) has become the *de facto* standard for many disciplines and on many academic campuses. **The NIH should raise the NRSA postdoctoral starting salary to $50,000 (2014 dollars), and adjust it annually for inflation. Postdoctoral salaries should be appropriately higher where regional cost of living, disciplinary norms, and institutional or sector salary scales dictate higher salaries.**[1]

In addition, host institutions should provide benefits to postdoctoral researchers that are appropriate to their level of experience and commensurate with benefits given to equivalent full-time employees. Comprehensive benefits should include health insurance, family and parental leave, and access to a retirement plan.

This recommendation requires action primarily by the funding agencies, with additional actions by the host institutions and the professional societies.

4.1 Federal agencies should require host institutions to provide documentation of the salary a postdoctoral researcher will receive with all grant proposals.

4.2 Professional societies should collect data on salaries for all positions and make these publicly available.

5. *Mentoring*: Mentoring is an essential component of the postdoctoral experience and entails more than simply supervision. Mentoring should not be solely a responsibility of the principal investigator, although he or she should be actively engaged in mentoring. **Host institutions should create provisions that encourage postdoctoral researchers to seek advice, either formally or informally, from multiple**

---

[1] Two of the committee members do not support the recommendation for a prescriptive "salary standard" based upon one particular field and funding agency (here, the National Institutes of Health [NIH] and life sciences) for two reasons: first, salaries—not just postdoctoral salaries—differ so much by discipline, region, funding agency, and type of institution (for example, the 2012 National Postdoctoral Association report indicates that about half of the institutions have minimum salaries that are lower than the 2013 NIH minimum of $39K; NPA 2012), and second, this "salary standard," meant to reflect a reasonable salary, will likely be used as a minimum salary. While they believe that institutions need flexibility to accommodate particular circumstances, they also firmly believe that a postdoctoral researcher's salary should be fair and fit rationally within the spectrum of salaries for researchers in that discipline, at that institution: for example, well above that of a graduate student and significantly less than that of an entry-level, career-track researcher, that is, permanent staff scientist, research track assistant professor, or tenure-track assistant professor.

advisors, in addition to their immediate supervisor. Host institutions and funding agencies should take responsibility for ensuring the quality of mentoring through evaluation of, and training programs for, the mentors.

This recommendation requires action by the funding agencies and the host institutions, with supporting actions by the professional societies, the mentors, and the postdoctoral researchers themselves.

5.1 In addition to providing mentorship training and guidance to the immediate supervisors of the postdoctoral researchers, host institutions should establish mechanisms that make it easy for postdoctoral researchers to seek guidance from additional faculty or senior professionals who can enrich the postdoctoral training experience.

5.2 Funding agencies should identify better ways of evaluating or rewarding mentoring as an essential component of research. This could include mandatory self-reporting by mentors as well as blinded assessments by the postdoctoral researchers.

5.3 Professional societies are in an ideal position to provide additional mentors to supplement those at a postdoctoral researcher's host institution. This would be of particular value to postdoctoral researchers considering major career shifts such as a move from academia to industry.

5.4 Postdoctoral researchers need to recognize that a great research investigator is not necessarily equivalent to a great mentor and that many if not most principal investigators or senior research faculty have not received any formal training in mentoring. Therefore, postdoctoral researchers should seek guidance from a variety of people, and should be encouraged to do so.

6. ***Data Collection***: Current data on the postdoctoral population, in terms of demographics, career aspirations, and career outcomes are neither adequate nor timely. Every institution that employs postdoctoral researchers should collect data on the number of currently employed postdoctoral researchers and where they go after completion of their research training, and should make this information publicly available. **The National Science Foundation should serve as the primary curator for establishing and updating a database system that tracks postdoctoral researchers, including non-academic and foreign-trained postdoctoral researchers.** Host institutions and federal agencies should cooperate with NSF on the data collection and maintenance process. Federal agencies and research institutions that report these data to the NSF should take advantage of various

technologies that have become available in recent years to assist in timely and thorough collection.

Recognizing that this recommendation on data collection has been made many times before with little effect, the committee stresses that research institutions and professional societies should explore what they can do to enrich what is known about postdoctoral researchers and that all institutions make better use of new technologies and social and professional networks to collect relevant and timely data.

This recommendation requires action primarily by the funding agencies, with additional actions by the host institutions and the professional societies.

6.1 Funding agencies must improve their data collection on the postdoctoral segment of the workforce. This is especially true for the NSF, given its congressional mandate to "collect, acquire, analyze, report, and disseminate statistical data related to the science and engineering enterprise in the United States and other nations that is relevant and useful to practitioners, researchers, policymakers, and the public, including statistical data on research and development trends, [and] the science and engineering workforce… " (Section 505 of the America COMPETES Reauthorization Act of 2010). The NSF should work with other research agencies, particularly the NIH, to develop more reliable means of collecting data on postdoctoral researchers during and after their appointments. The use of a common identifier system for each postdoctoral researcher is a possible approach.

6.2 Host institutions should assist in the data collection efforts by remaining consistent with their labeling of postdoctoral researcher, keeping track of new hires and departures, and conducting exit interviews to determine career outcomes of their postdoctoral population. This information should be made publically available, particularly to prospective postdoctoral researchers.

6.3 Funding agencies should look favorably on grant proposals that include outcome data for an institution's postdoctoral researchers.

6.4 Professional societies should utilize their networks to collect information about career paths of their members and make this data easily available.

All of the reforms recommended here should be coordinated through a strong and separate or stand-alone postdoctoral office (PDO) at each host institution. These offices have become much more common since the publication of the 2000 Postdoctoral Report, and many have become members of the National Postdoctoral Association. However, more work is needed to truly enrich the postdoctoral experience. PDOs need to continue sharing experiences to help one another fulfill their potential to train mentors, organize career development activities, be a one-stop source of information for domestic and international postdoctoral researchers, manage postdoctoral researcher grievances, oversee data-gathering efforts, monitor institutional compliance with salary and benefits policy, and track the career progress of former postdoctoral researchers. Although currently these offices are often embedded within a larger graduate student affairs operation, they are essential for improving the visibility and recognition of postdoctoral researchers in their host institutions and deserve specialized recognition.

A larger goal of this study was not only to propose ways to make the postdoctoral system better for the postdoctoral researchers themselves but also to better understand the role that postdoctoral training plays in the research enterprise. The committee asked whether there are alternative ways to satisfy some of the research and career development needs of postdoctoral researchers that are now being met with several years of advanced training. The committee hopes that this report stimulates action toward clarifying the role of postdoctoral researchers in the research enterprise, and improving their status and experience.

1
Introduction

Concern about postdoctoral training is not new. In 1969, Richard B. Curtis wrote the following in the preface to the National Research Council's (NRC) *Invisible University: Postdoctoral Education in the United States*:

> [T]he postdoctoral phenomenon needs study just because it has been so successful. Increasing numbers of postdoctoral students have caused them to become visible beyond the laboratory and the library. But it would be more accurate to say that the larger community has become aware of them without really seeing them. (NRC 1969)

Although that report was, on the whole, a positive review of the use of postdoctoral researchers, in 1981 the National Research Council report *Postdoctoral Appointments and Disappointments* (NRC 1981) identified serious concerns about the postdoctoral enterprise (Box 1-1) and concluded by calling for a broader reevaluation of the enterprise.

Two decades later, the Committee on Science, Engineering, and Public Policy (COSEPUP) oversaw another report, *Enhancing the Postdoctoral Experience for Scientists and Engineers* (the 2000 Postdoctoral Report). It stressed that a rapid expansion of postdoctoral training had taken place without adequate oversight, resulting in fundamental changes in the nature of the experience for many postdoctoral researchers. To improve the postdoctoral system, it offered specific directions for a range of stakeholders: federal agencies, universities, foundations, professional organizations, and postdoctoral researchers themselves.

Box 1-1
***Postdoctoral Appointments and Disappointments* (1981) Key Issues**

1. The lack of prestige and research independence in postdoctoral appointments for the most talented young people;
2. The mismatch between the important role that postdoctorals play in the nation's research enterprise and the lack of opportunities that they find for subsequent careers in research; and
3. The lack of recognized status of postdoctoral appointments in the academic community.

SOURCES: NRC 1981

The 2000 Postdoctoral Report identified 3 principles to guide postdoctoral training, and 10 actions that should be taken by advisers, institutions, funding organizations, and disciplinary societies to improve the postdoctoral training system (Box 1-2).

Aware of the continuing concerns regarding employment compensation, benefits, and length of time to transition from postdoctoral researchers into permanent positions, in 2011 the National Academies formed an *ad hoc* committee under the auspices of COSEPUP to review the state of the postdoctoral experience. As part of its task (Box 1-3), the committee was to determine whether the recommendations made in the 2000 Postdoctoral Report

Box 1-2
The 2000 Postdoctoral Report

Principles

1. The postdoctoral experience is first and foremost a period of apprenticeship for the purpose of gaining scientific, technical, and professional skills that advance the professional career.
2. Postdocs should receive appropriate recognition (including lead author credit) and compensation (including health insurance and other fringe benefits) for the contributions they make to the research enterprise.
3. To ensure that postdoctoral appointments are beneficial to all concerned, all parties to the appointments—the postdoc, the postdoc adviser, the host institution, and funding organizations—should have a clear and mutually-agreed-upon understanding with regard to the nature and purpose of the appointment.

Actions

1. Award institutional recognition, status, and compensation commensurate with the contributions of postdocs to the research enterprise.
2. Develop distinct policies and standards for postdocs, modeled on those available for graduate students and faculty.
3. Develop mechanisms for frequent and regular communication between postdocs and their advisers, institutions, funding organizations, and disciplinary societies.
4. Monitor and provide formal evaluations (at least annually) of the performance of postdocs.
5. Ensure that all postdocs have access to health insurance, regardless of funding source, and to institutional services.
6. Set limits for total time of a postdoc appointment (of approximately five years, summing time at all institutions), with clearly described exceptions as appropriate.
7. Invite the participation of postdocs when creating standards, definitions, and conditions for appointments.
8. Provide substantive career guidance to improve postdocs' ability to prepare for regular employment.
9. Improve the quality of data both for postdoctoral working conditions and for the population of postdocs in relation to employment prospects in research.
10. Take steps to improve the transition of postdocs to regular career positions.

SOURCES: The 2000 Postdoctoral Report

Box 1-3
Statement of Task

Building on the 2000 COSEPUP report *Enhancing the Postdoctoral Experience for Scientists and Engineers*, an *ad hoc* committee will describe the state of postdoctoral programs in the United States, examine how postdoctoral fellows (postdocs) are being guided and managed, review institutional practices with respect to postdocs, try to determine what happens to postdocs after they complete their programs, explore important changes that have occurred in the postdoc practices and in the research ecosystem, and assess how well current practices meet the needs of these fledgling scientists and engineers and of the research enterprise.

Based on a review of existing data about postdocs and institutional practices, the committee will, to the extent possible, attempt to answer key questions in the following areas:

1. **General characteristics of postdoctoral fellows and positions in the United States:** How many postdoctoral fellows are there in the United States? Where are they working, in what fields, and for how many years?
2. **Current conditions for postdocs:** Are expectations of principal investigators made clear? Do postdocs receive adequate professional status and privileges as well as salary and benefits? Are the rules clear about credit they receive for their discoveries in the lab, and are they receiving adequate career guidance and development?
3. **Institutional provisions:** Do postdocs serve as investigators on grants? Are questions of intellectual property identified and provided for? At universities, is teaching required; if not, is it encouraged or discouraged?
4. **Career paths:** Where do postdocs come from? What do we know and what can we learn about what postdocs do after they complete their programs. How well are the postdoc programs matched with the career opportunities that are open to them?
5. **Recent trends and changes:** Have previous recommendations been implemented and to what effect? Are there other developments in the research enterprise that have had a significant effect on postdocs?
6. **Participation in the research enterprise:** Are postdocs being invited to review journal articles and to write grant proposals, either formally by journals and agencies or informally by principal investigators, and is this experience useful? What are the impressions of postdocs about peer review today? Are postdocs being used effectively in research? Are postdocs acquiring the skills they need to become productive independent researchers in the future?

had been implemented, whether the conditions of postdoctoral training had changed, and whether there was a need to consider further actions to improve the postdoctoral experience in light of a dynamic and continuously changing research enterprise. The committee reviewed the available data from the National Science Foundation (NSF) and other sources; heard testimony from numerous postdoctoral researchers; met with senior officials of NSF, the National Institutes of Health (NIH), and international research programs; and spoke with leaders of a variety of research institutions. The committee examined the status of postdoctoral training as well as questions of how postdoctoral researchers fit into the larger structure of the research enterprise, with the goal of not merely making the best of the existing system, but trying to consider how well the system meets the needs of aspiring researchers and the entire research establishment, and how it might be modified if it was not optimal.

Some of the major findings from this examination have become more common topics of discussion and concern. The number of postdoctoral researchers in science, engineering, and health has increased dramatically, up nearly 150 percent between 2000 and 2012. That far surpasses both the percentage increases in graduate students and in tenure and tenure track faculty positions over the same time period. The demographics of postdoctoral researcher are changing as more women and noncitizens are entering this segment of the research workforce. In response to all this growth, a number of institutional structures, at many levels, have developed over the past decade and a half, but the actions taken have not been adequate.

A significant frustration for anyone trying to understand the postdoctoral system is the paucity of comprehensive data. One of the key findings of the 2000 Postdoctoral Report was that the data did not exist to construct a complete picture of postdoctoral training. Unfortunately, in the nearly 15 years since that observation, the picture has still not been completed. The most consistent sources of data on postdoctoral researchers are the several surveys of graduate training conducted by the NSF, some of which are run in conjunction with the NIH. Three current surveys in particular target postdoctoral researchers: Survey of Graduate Students and Postdoctorates in Science and Engineering (GSS), the Survey of Doctorate Recipients (SDR), and the Survey of Earned Doctorates (SED).

Each source of data has clear weaknesses regarding postdoctoral researchers. The 2008 edition of NSF's *Science and Engineering Indicators* included a long discussion of postdoctoral training that acknowledged the limitations of its data collection (NSB 2008). Of significance to this current report was the following finding:

> No single data source measures the entire population of postdocs, and some parts of the population are not systematically measured at all. Two NSF surveys, the Survey of Doctorate Recipients (SDR) and the Survey of Graduate Students and Postdoctorates in Science and Engineering (GSS), include data bearing on the number of postdocs in the United States.
>
> SDR covers U.S. residents who have earned [science and engineering] and health doctorates from U.S. schools (MDs and other types of degrees with "doctor" in the name are not included). Thus, postdocs who received doctorate degrees from foreign institutions are not included in SDR. In 2006, SDR collected data on the dates of current and past postdoc positions, allowing an

> estimate to be made of the number of postdocs in fall 2005, the same period as the most recent GSS data. Unlike SDR, which collects data from individuals, GSS surveys academic departments. GSS asks departments that offer graduate programs in [science and engineering] and specific health-related fields for counts of all of their postdocs, regardless of whether their degrees were earned in the United States or abroad. However, unlike SDR, it does not gather data on people in nonacademic positions or academic units that lack graduate programs, including many academic research organizations and affiliated nonprofit research centers.

The SDR, which tracks where people go after completing doctoral training, does not include postdoctoral researchers who earned their degrees in other countries, even though more than half of all postdoctoral researchers currently working in the United States fall into this category. The GSS surveys institutions, not individuals, and these institutions often find it difficult to identify everyone who should be counted, although recent methodological improvements are working toward rectifying some of the suspected under-counting (Einaudi 2013).[2] Although the GSS does include those who earned Ph.D.'s in other countries, it primarily surveys only those departments that offer graduate programs. Many postdoctoral researchers work in research institutes, national laboratories, and companies that do not offer graduate training, and institutions can differ in whom they define as a postdoctoral researcher.

The third survey, the SED, is used to determine what percentage of Ph.D.'s plan to pursue postdoctoral training, and shares the same sampling frame as the SDR. It asks people receiving Ph.D.'s from U.S. institutions what they expect to be doing after graduation, but definite commitments are only reported for approximately two-thirds of the surveyed population. Of these, there is additional, albeit perhaps small, uncertainty that this is what they will actually pursue as their stated postgraduation positions. Therefore, the SED data on U.S.-degreed Ph.D. recipients with definite commitments for postdoctoral researchers offers only a very limited snapshot of the incoming cohort of postdoctoral researchers in a given year.

Finally, lack of timeliness continues to be a major drawback of NSF data. Results from some of the 2009 surveys did not appear until late 2013. When one considers the enormous economic upheaval the nation has experienced since 2008, it is obvious these data have limited value in understanding what was happening during this tumultuous period.

---

[2] Note from the GSS: "In 2010, the postdoc section of the survey was expanded, and significant effort was made to ensure that appropriate personnel were providing postdoc data … . Thus, for increases in 2010 or 2011 over 2009 and prior-year data, it is unclear how much is from growth in postdoctoral appointment and how much is from improved data collection." Details about the methodical changes and resulting implications are covered in Einaudi 2013.

With all the limitations described above, the committee had to adopt a policy regarding the use of the available data. It has little confidence in the accuracy of the absolute number of postdoctoral researchers, and it is particularly dubious about the quality of the information about postdoctoral researchers who are temporary residents and earned their Ph.D.'s in other countries. Nevertheless, the committee considers the available data to be a reliable indicator of trends over time. The gaps and flaws that exist are the same gaps and flaws that have existed for decades, so at least it may be supposed that the data possess some internal consistency.

At critical points in the report the committee reminds the reader that some data need to be understood in context and hopes that throughout the report the data presented are the best available, but clearly all data must be interpreted with the caveats discussed above.

The committee also needs to comment on its decision not to address the final item in its statement of task (see Box 1-3): participation in the research enterprise. Although the committee recognizes that it is important to know the extent to which postdoctoral researchers are gaining experience in reviewing journal submissions and writing grants, there is simply no way to accurately acquire that information. These are questions that ideally would be asked in surveys of postdoctoral researchers. Similarly, it is of critical importance to know whether postdoctoral researchers are being used effectively in research and acquiring the skills they need to advance their careers; however, no one is collecting that information or evaluating such practices at research institutions. The committee concluded that it could not provide an informed answer to these questions.

2
The Disconnect Between the Ideal and Reality

Research training immediately after the doctoral degree can be a productive period for many researchers, and an essential step on the path to an independent research career. The experiences gained during these postdoctoral years facilitate the development of research careers in science, engineering, and medicine. Unfortunately, the term "postdoc"—as the period of postdoctoral training and the people who purse postdoctoral training have both come to be colloquially known—is associated with many interpretations (see Box 2-1), and too often does not align with the ideal of a temporary and definite period of mentored advanced training in research.

The standard model of academic scientific research that has developed in the United States over the past century is based on a core research group led by at least one principal investigator, and may include staff scientists, postdoctoral researchers, graduate students, technicians, and even undergraduates. In this structured research unit, postdoctoral researchers play a critical role in supporting the day-to-day operations of a research laboratory. In return, they gain experience to further their own independent research careers (Stephan 2012).

The ideal scenario for a postdoctoral researcher's experience might be as follows.[3] A self-selected, motivated, recent doctoral degree recipient receives additional scientific training (to augment previous training during graduate school) for a limited time before transitioning to a full-time research position, often as a tenure-track faculty member. Under the mentorship of a principal investigator, the postdoctoral researcher learns new tools and techniques of research and gains access to professional networks and relationships that are crucial to securing a faculty or other position. Along the way, the postdoctoral researcher accrues publications—the valued currency of the academic and research worlds—and learns the arcana of grantsmanship and other vital skills of the trade.

In exchange for the skills and knowledge gained during this training, the postdoctoral researcher provides highly skilled, low-cost research labor to the laboratory, often bringing new techniques that complement the principal

[3] This is primarily for university-based postdoctoral researchers, which is the largest category.

Box 2-1
What is in the Name "Postdoc"?

The nomenclature that is used to refer to this increasing segment of our science and engineering workforce is reflective of many of the issues that arise in this report. The word "postdoc" is a truncation of the adjective "postdoctoral," which describes a period of time, and is conveniently used to modify many types of positions, including traineeships, fellowships, research scientists, and other related titles. In referring to people who are in postdoctoral positions, the word "postdoc" has also become a noun.

A "postdoc" is the only stage in the education, training, and career development of a scientist or engineer that is referred to only by the period of time, and not the position itself. Because there is no agreed-upon specific title, there is no agreed-upon set of goals or outcomes for the position. Is it intended to be a continuation and enhancement of the in-depth education of the Ph.D. and graduate school (traineeship)? Or is it an entry-level career position with the corresponding benefits and job training expected from any entry-level position (researcher)?

In addition, by referring to the time period and not the position, there is little consistency in how institutions identify individuals during this stage or how individuals identify themselves. This conundrum hampers data collection efforts. Some—perhaps, many—of those individuals currently labeled "postdocs" by their principal investigators, institutions, or funding agencies should hold very different titles (e.g., research scientist).

Even the time period that the adjective "postdoctoral" describes is ambiguous and ill-defined. The only qualification that is required for any position to be postdoctoral is that it occurs after the receipt of a doctoral degree (in the European Union, it can be argued to be after the receipt of any terminal degree). This period is open ended and therefore, taken glibly, once doctoral candidates have completed their degree, they are technically, and indefinitely, postdoctoral. Practically, this term stops being used once they transition to a different role, like professor or research scientist.

In this report, the committee determined that the most appropriate title for this segment of the workforce is "Postdoctoral Researcher."

investigator's research group. The postdoctoral researcher contributes directly to publications and grants—and hence to the productivity of the research group. The postdoctoral researcher may also work with graduate students and other laboratory members, allowing the principal investigator to attend to duties such as teaching, fundraising, and administration. Ideally, this symbiotic relationship benefits all parties involved.

HISTORICAL PERSPECTIVE

During the first half of the 20th century, when the scale of scientific research was relatively small, the reality of the postdoctoral experience was strongly aligned with the ideal (see Box 2-2). However, the dramatic shift toward "big science" during and following World War II, which has only increased in the subsequent decades, has resulted in a system that often requires ever larger teams of graduate students, postdoctoral researchers, and other

Box 2-2
Historical Context of the Postdoctoral Enterprise in the United States

The Start: The current system of postdoctoral research in the United States arose after World War I in an effort led by the National Research Council and the Rockefeller Foundation to strengthen basic research in physics and chemistry to be competitive with other leading nations, especially Germany. In the 1920s and 1930s, postdoctoral researchers quickly developed a role in managing the day-to-day research operations of laboratories, allowing faculty time to supervise other research, teach, obtain funding, and attend to other administrative tasks.[a]

The Surge: Following World War II, funding for basic research at universities shifted from foundations and private industry to the federal government. The growth in the numbers of postdoctoral researchers mirrored the general increase in funding. In the 1960s, approximately 10 percent of doctoral graduates went on to a postdoctoral position, although the growth in the biomedical fields was quickly outpacing other areas of science and engineering.[b] Initially, most federal grants were awarded to individual faculty members at universities and covered only research expenses. However, indirect rates were increased rapidly, and soon salaries, including those for postdoctoral researchers, were included in research grants.[c]

The Boom: The ranks of postdoctoral researchers grew even more steeply beginning in the 1970s, as did the average duration of postdoctoral appointments. The selection of postdoctoral researchers switched from the "cream of the crop" who won competitive fellowships to a period in which the majority of postdoctoral researchers were supported as research assistants.[d]

The Maturity: By the end of the 20th century, the postdoctoral research position had become an established component of professional training in many fields of science and engineering, and was becoming more common in the social sciences and humanities. The position still retained its essential attributes of providing advanced research training and assisting with the research operations of laboratories. Although the value of postdoctoral researchers to the conduct of research remains clear, the value of this experience became questionable for many postdoctoral researchers.[e]

SOURCES:

[a] Assmus, Alexi. "The Creation of Postdoctoral Fellowships and the Siting of American Scientific Research." *Minerva* 31, no. 2 (1993): 151-83.

[b] National Research Council. "Invisible University: Postdoctoral Education in the United States. Report of a Study Conducted Under the Auspices of the National Research Council. [Richard B. Curtis, Study Director]." Washington, DC: The National Academies Press, 1969.

[c] Stephan, Paula. "The Endless Frontier: Reaping What Bush Sowed?" NBER Working Paper 19687. Available at http://www.nber.org/papers/w19687, accessed on April 10, 2014.

[d] Ibid.

[e] Alberts, Bruce, Marc W. Kirschner, Shirley Tilghman, and Harold Varmus. "Rescuing US biomedical research from its systemic flaws," *PNAS* (Perspective) 111(16): 5773-5777 April 22, 2014. doi:10.1073/pnas.1404402111

researchers working under a principal investigator. While the demand for junior research workers has boomed, the number of research faculty positions into which the junior researchers could hope to move has not kept pace, especially in academia, and tenured faculty are not retiring at the rate that their aspiring replacements are being trained. The result is a system that has created expectations for academic career advancement that cannot be met.

Whereas the number of postdoctoral fellowships and traineeships (which are designed to meet the development needs of new Ph.D.'s) has remained constant, the number of postdoctoral positions created as part of research grants has grown steadily (see Figure 2-1). Principal investigators were aware that many of their postdoctoral researchers were not able to find independent research or faculty positions, but it was not obvious what could be done about it. Many principal investigators, trained in the post–World War II system, had relatively little experience outside of academia and therefore lacked the direct knowledge of other sectors necessary to provide quality mentoring to the growing number of students and postdoctoral researchers who would end up in career paths other than independent, academic research.

A BASIC DEFINITION

By the mid-2000s, the need to formalize a definition of the postdoctoral researcher arose. The NIH, NSF, Association of American Medical Colleges

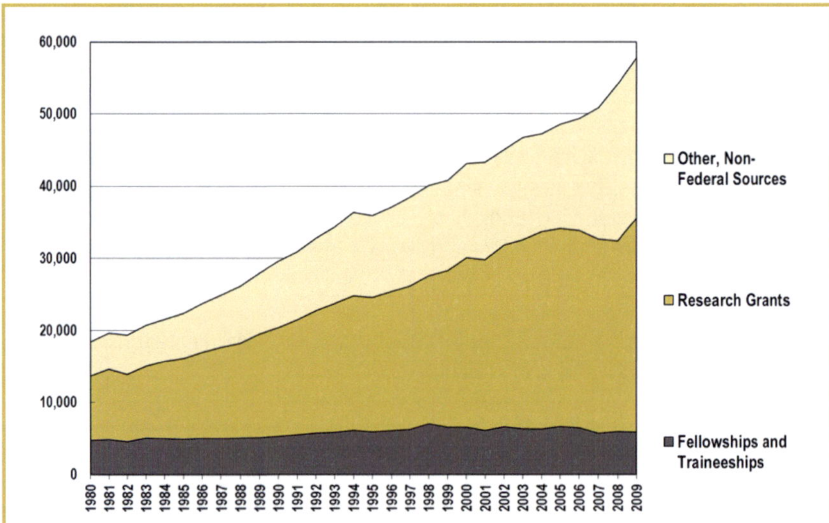

FIGURE 2-1 Variations in types of postdoctoral researchers working at academic institutions with graduate programs by mechanism of support.
NOTE: Includes science, engineering, and health postdoctoral researchers.
SOURCE: Data compiled from the NSF and NIH Graduate Students and Postdoctorates in Science and Engineering Survey (GSS) via WebCaspar.

(AAMC), and National Postdoctoral Association (NPA) all developed definitions (see Box 2-3). All converge on three basic aspects: (1) prior completion of a doctoral degree; (2) a primary focus on advanced research training (that includes publications, grants, and professional networking); and (3) a fixed term of appointment.[4]

THE CURRENT REALITY

Although the experience of some postdoctoral researchers does come close to the ideal principles and definitions, increasingly the current reality differs—sometimes in extreme ways. One factor driving this divergence is the sharp increase in the number of postdoctoral researchers. Fewer recent U.S.-trained doctorates have definite commitments now than at the time of the publication of

Box 2-3
Postdoctoral Researcher: Definitions and Characterizations

- **National Institutes of Health and National Science Foundation (2007):** "An individual who has received a doctoral degree (or equivalent) and is engaged in a temporary and defined period of mentored advanced training to enhance the professional skills and research independence needed to pursue his or her chosen career path."
- **National Postdoctoral Association (2007):** "A postdoctoral scholar ('postdoc') is an individual holding a doctoral degree who is engaged in a temporary period of mentored research and/or scholarly training for the purpose of acquiring the professional skills needed to pursue a career path of his or her choosing."
- **Association of American Medical Colleges (2006):** "Postdoctoral training is an integral component of the preparation of scientists for career advancement as scientific professionals. Postdoctoral appointees typically join an institution to further their training in a chosen discipline after recently obtaining their terminal degree (e.g., Ph.D., MD, DVM). This training is conducted in an apprenticeship mode where she/he works under the supervision of an investigator who is qualified to fulfill the responsibilities of a mentor. The postdoctoral appointee may undertake scholarship, research, service, and teaching activities that together provide a training experience essential for career advancement."

SOURCES:
NSF: The National Science Foundation. "Proposal and Award Policies and Procedures Guide: Part I - Proposal Preparation & Submission Guidelines (GPG)." Effective June 1, 2007. NSF 07-140. OMB Control Number: 3145-0058. April 2007.
NIH: Available at http://grants1.nih.gov/training/q&a.htm. Accessed May 8, 2014
NPA: Available at http://www.nationalpostdoc.org/social-media-terms. Accessed May 8, 2014
AAMC: Association of American Medical Colleges. "Compact between Postdoctoral Appointees and Their Mentors." Washington, DC, 2006.

---

[4] The period of postdoctoral appointments is highly variable, with terms ranging from a year to a decade. It should be noted that, whereas the definitions used by the NIH, NSF, and NPA all state explicitly that a postdoctoral position is a "temporary" position, the AAMC definition only does so implicitly by referring to it as "training conducted in an apprenticeship mode," (see Box 2-3).

the 2000 Postdoctoral Report, regardless of field,[5] and those that do are more likely to declare that they are entering into a postdoctoral research position. Nearly 40 percent of all 2012 doctorate recipients with postgraduate plans were committed to postdoctoral study, according to the most recent NSF SED data, down slightly from its all-time high of more than 43 percent in 2010[6] when *American Recovery and Reinvestment Act* funding for postdoctoral scholars was available. Limiting the data to just life science, physical science, social science, and engineering, the rate increases to approximately 50 percent in 2012.[7] In the United States alone, an estimated 60,000 to 100,000 postdoctoral researchers work in various research fields. According to the GSS, postdoctoral researchers in the biomedical sciences constitute the largest fraction of the current U.S. postdoctoral population, followed by those in the physical sciences (see Figure 2-2). Globally, it is estimated that the number of postdoctoral researchers may be anywhere between 250,000 and 400,000 (Gallagher 2012).

There is a wide variation in the experience of postdoctoral researchers—a variation that is due to the diversity of practices in the scientific enterprise. In fact, there appears to be no standard practice in the ways in which postdoctoral researchers are hired, mentored, and transitioned.

Location

Most postdoctoral researchers are based in universities and other research centers. Universities obviously cannot provide jobs for all of their postdoctoral researchers at the end of their terms, and the committee members agreed that in their experience, at a wide variety of universities, the expectation is that the postdoctoral researchers will find jobs elsewhere. In contrast, private companies and national laboratories often view the postdoctoral period as an opportunity to observe potential employees, and many hire their own postdoctoral researchers into more permanent positions at the end of their research contracts.

Most discussions of postdoctoral training refer to postdoctoral researchers working at universities, but approximately 11 percent work at federally funded research and development centers (FFRDCs) such as national laboratories, or in industry research and development centers.[8] Testimony from current postdoctoral researchers and the experience of some committee members

[5] Aggregate across all disciplines, between 2000 and 2012, the percentage of recent U.S.-trained doctorates with definite commitments has fallen from above 70 percent to around 65 percent. For recent U.S.-trained doctorates in only life science, physical science, social science and engineering, the decrease has been from nearly 60 percent to approximately 51 percent.

[6] For a detailed discussion regarding sources of data on the postdoctoral population, please see Chapter 1.

[7] In the SED data life sciences include agricultural sciences, natural resources, biological and biomedical sciences, and health sciences; physical sciences include mathematics and computer and information sciences; social sciences include psychology.

[8] The 11 percent figure is derived from a combination of NSF survey data (SED, SDR, GSS, and the FFRDC postdoc surveys).

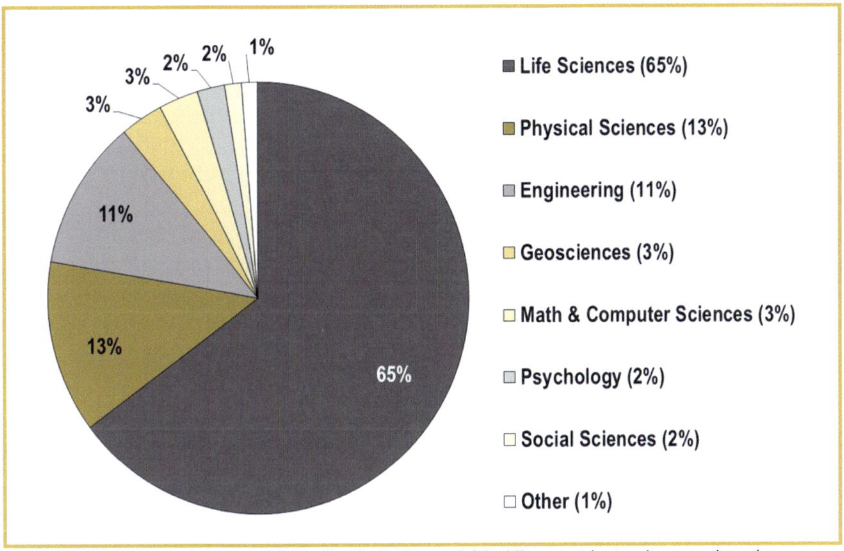

FIGURE 2-2 Doctoral degree (Ph.D., Sc.D., D.Eng. only) holding postdoctoral researchers in academia as reported by host institutions, by field: 2012.
NOTES: In fall 2012, GSS data were collected from 13,952 departments or organizational units at 367 SEH doctorate-granting institutions and 198 SEH master's-granting institutions, for data on 43,426 doctoral degree (Ph.D., Sc.D., D.Eng. only) holding postdoctoral researchers. "Other" includes Architecture and Environmental Design, Communication and Librarianship, Interdisciplinary or Other Sciences, and Vocational Studies and Home Economics.
SOURCE: Data compiled from the NSF and NIH Graduate Students and Postdoctorates in Science and Engineering Survey (GSS) via WebCaspar.

indicated that the experience of these postdoctoral researchers differs in significant ways from their academic counterparts: salaries are higher, the term of the appointment is usually shorter, their position in the institution is more clearly defined, and there is a reasonable chance that the postdoctoral researcher will eventually be hired for permanent employment. For the FFRDCs, postdoctoral fellows are increasing in importance for attraction of top scientific talent. For this reason, and motivated by feedback from postdoctoral researchers, the FFRDCs are currently collaborating to increase the quality of research experience and mentoring received by postdoctoral researchers.

Type

Funding for postdoctoral research positions at universities commonly falls into three main categories: fellowships, training grants, and research grants. Currently, the vast majority fall into the latter category. Figure 2-1 (above) shows that the number of postdoctoral researchers on fellowships and traineeships has remained fairly constant over the past several decades. However, there has been notable growth in the number of postdoctoral researchers supported by research grants and those supported by nongovernmental sources, including foundations and other governments.

Postdoctoral researchers with some independent competitive fellowships often have at least some direct control over their research topic and can typically choose where to study and to relocate, if desired. Such portability is an advantage. Typically, training grants are awarded to research institutions, and postdoctoral researchers are selected by the grant-receiving institution, but when the postdoctoral researcher holds the traineeship, she or he may retain some flexibility in choice of research topic. The most common type of funding mechanism is through the principal investigators, who recruit postdoctoral researchers directly to work on research grant awards they have been able to secure from federal or private sources. In this model, postdoctoral researchers work under the direct supervision of principal investigators in an area that is of interest to their host laboratories, and they have little mobility.

Field

Postdoctoral experiences also vary significantly by research fields. Many of these differences are due to specific disciplinary factors, such as the time it takes to produce data from research experiments and publish in peer-reviewed journals, the availability of funding for various research areas, and the availability of jobs that require postdoctoral research experience.

Over the past two decades, the percentage of Ph.D. recipients who entered postdoctoral positions grew in every discipline (see Figure 2-3), with the rate of growth fastest in those disciplines where postdoctoral training was relatively rare in 2000, such as engineering (up 13.3 percentage points between 2000 and 2012) and the social sciences (up 12.7 percentage points between 2000 and 2012). The most recent Survey of Earned Doctorates (SED) data for Ph.D. recipients who earned degrees at U.S. institutions in 2012 indicate that more than 65 percent of life scientists, almost 55 percent of physical scientists, nearly 35 percent of engineers, and more than 36 percent of social scientists with definite commitments were entering postdoctoral training.[9] Ph.D.'s in the biomedical sciences have traditionally been the most likely to pursue postdoctoral training, with rates above 60 percent for more than a decade. Excluding the humanities, the social sciences and engineering have the lowest percentage of U.S.-degreed Ph.D.'s entering postdoctoral training. The physical sciences fall in the middle for rates of postdoctoral participation. In terms of duration, a single postdoctoral appointment still lasts approximately 2 years (2006 Survey of Doctorate Recipients [SDR] data), but in total, an individual

[9] In 2012, the percentage of recent U.S.-trained doctorates with definite commitments in the life sciences was just over 62 percent, physical sciences was just over 68 percent, engineering was approximately 64 percent, and social sciences was nearly 70 percent.

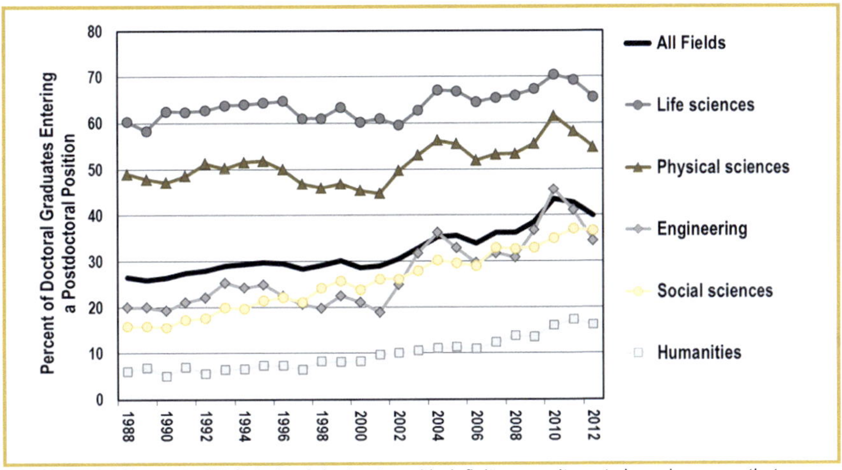

FIGURE 2-3 Percent of U.S.-trained doctorates with definite commitments in a given year that go on to postdoctoral positions that year by broad fields of research.
SOURCE: Data compiled from the NSF, NIH, U.S. Department of Education, U.S. Department of Agriculture, National Endowment for the Humanities, and NASA Survey of Earned Doctorates (SED): 2012, 2011, 2010, 2009, and 2008.

spends a median of 3 to 4 years as a postdoctoral researcher.[10] It is not unusual to find biomedical researchers who have completed several postdoctoral appointments that total more than 5 years.

The employment of postdoctoral researchers varies by field as well. According to the SDR, which follows only U.S.-degreed Ph.D. recipients, the life and physical sciences have the highest percentage of postdoctoral researchers in the academic workforce—both more than 10 percent of the total workforce—and the correspondingly lowest percentage of tenure and tenure-track faculty (see Figure 2-4). It is important to note that this most likely is a dramatic undercounting of the use of postdoctoral researchers, caused by the limited sampling frame of the survey.

Postdoctoral salaries cover a wide range; Table 2-1 compares median annual salaries for individuals who recently received their science, engineering,

[10] Kahn 2011 and from the supplemental material of the NIH Biomedical Workforce Report: "Median length of time with the title postdoc, estimated from the SDR. Because the SDR only covers US-trained doctorates and because foreign doctorates may do longer postdocs, this number may be an underestimate. For comparison purposes, with 8700 postdoctoral entrants per year and 68,000 total postdocs, we would expect a mean postdoc length of 7.8 years (due to the distribution being right-skewed, we do expect the mean to be larger than the median)." Available at http://report.nih.gov/investigators_and_trainees/ACD_BWF/PhD_Postdoctoral.aspx. Accessed July 7, 2014.

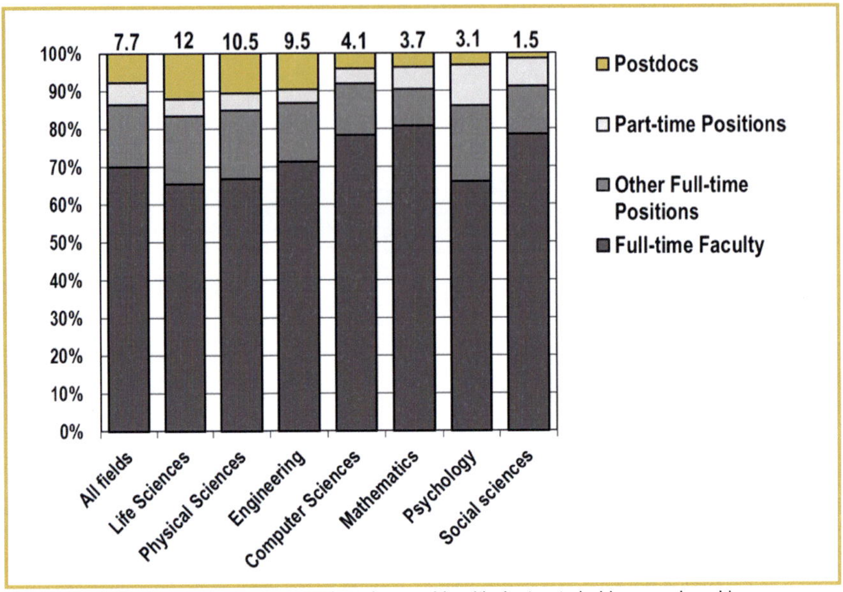

FIGURE 2-4 U.S.-trained science, engineering, and health doctorate holders employed in academia, by type of position and degree field, 2010.
NOTES: "Academic employment" is limited to U.S. doctorate holders employed at 2- or 4-year colleges or universities, medical schools, and university-based research institutes. "Physical sciences" include earth, atmospheric, and ocean sciences; "life sciences" include biological, agricultural, environmental, and health sciences. "Full-time faculty" includes full-time professors, full-time associate professors, and full-time junior faculty. "Other full-time positions" include research associates, adjunct positions, lecturers, instructors, and administrative positions. "Part-time positions" exclude those employed part time because they are students or retired.
SOURCE: Science and Engineering Indicators 2014; NSF, NCSES, special tabulations (2013) of the SDR (various years).

or health doctorates in the United States and had definite postgraduation commitments in employment or postdoctoral study. Postdoctoral researchers in some disciplines, at some institutions, and in some highly competitive fellowship positions earn salaries above $80,000 per year,[11] but median salaries are considerably lower. Beginning postdoctoral researchers in mathematics and computer science, business management, and economics reported median salaries above $56,000 a year in 2012, but the median for all postdoctoral researchers within five year of receiving their degree was only around $43,000

---

[11] NSF, NIH, U.S. Department of Education, U.S. Department of Agriculture (USDA), National Endowment for the Humanities, NASA, SED, 2012.

TABLE 2-1 Median salaries for recent U.S. science, engineering, and health (SHE) doctorate recipients in postdoc and non-postdoc positions up to 5 years after receiving degree: 2010 (Dollars)

Field of doctorate	Non-postdoctoral position	Postdoctoral position
All SEH	76,000	43,000
Life sciences[a]	65,000	42,000
Physical sciences	76,000	44,000
Computer & information sciences	97,000	48,000
Mathematics & statistics	70,000	53,000
Psychology	64,000	43,000
Social sciences	64,000	44,000
Engineering	91,000	44,000
Health	77,000	47,000

[a] Includes Biological, agricultural, and environmental life sciences.
NOTE: Salaries are rounded to the nearest $1,000. Data include graduates from 15 months to 60 months prior to the survey reference date.
SOURCE: National Science Foundation, National Center for Science and Engineering Statistics, Survey of Doctorate Recipients (2010), http://sestat.nsf.gov. Science and Engineering Indicators 2014

in 2010. These salaries are considerably lower than what their peers who pursued regular employment were earning.

Funding Source

The bulk of the federal funding for science, engineering, and health postdoctoral researchers in the United States comes from the NIH (~ 60 percent of federally supported postdoctoral researchers at academic institutions) and NSF (~12 percent of federally supported postdoctoral researchers at academic institutions). NIH funding is focused on biomedical and biomedical-related sciences; NSF funds research across a broad spectrum of physical sciences, engineering, social sciences, and some non-medical life sciences. Other federal agencies providing postdoctoral funding include the Department of Defense, the Department of Energy, NASA, and USDA.

Nonfederal funding also plays a role in postdoctoral researcher support, with more than 36 percent of postdoctoral researchers for whom sources were known either self-supporting or receiving nonfederal funding. Ranked by percentage of the total number of science, engineering, and health postdoctoral researchers at academic institutions supported (both federal and nonfederal), the reported sources are as follows: nonfederal–institutional (18–19 percent), nonfederal–domestic (13–14 percent), unknown/not reported (~7 percent), nonfederal–foreign (~3 percent), and self-support (<1 percent) (see Figure 2-5).

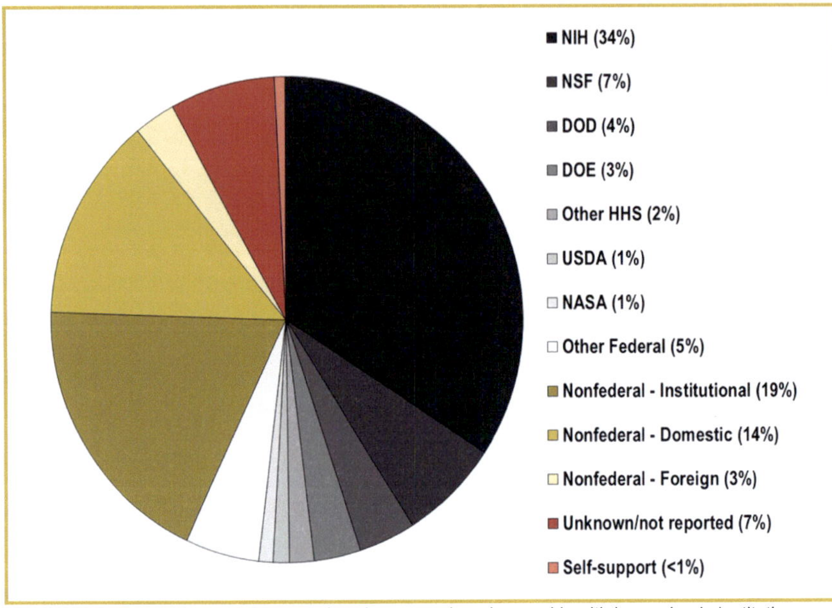

FIGURE 2-5 Postdoctoral appointees in science, engineering, and health in academic institutions, by field and primary source of support: 2012.
NOTES: Details on primary mechanism and source of support for postdoctoral researchers were collected for the first time in 2010. Percentages may not total to 100 percent due to rounding.
SOURCE: Data compiled from the NSF and NIH Graduate Students and Postdoctorates in Science and Engineering Survey (GSS) via WebCaspar.

DEMOGRAPHIC SHIFTS

The demographics of the postdoctoral population have changed in recent years. The percentage of women, temporary residents, and underrepresented minorities among postdoctoral researchers has continued to increase.

The beginning of this report includes a discussion of the committee's approach to the available data that is particularly relevant to this discussion of demographics.[12] The SED data includes detail on ethnicity, but does not include postdoctoral researchers who earned their degrees outside the United States. The GSS data include postdoctoral researchers with foreign Ph.D.'s, but does not collect data from some types of institutions that hire postdoctoral researchers. Additionally, international postdoctoral researchers may have funding from their home country and are, therefore, not included in the calculations of NIH, NSF,

[12] Further discussion in Chapter 1.

and other U.S. funders.[13] As a result, the committee finds only enough reliable evidence to identify a few significant trends in the postdoctoral population that are consistent across several datasets. They are summarized below, in percentages that are not meant to be precise.

International Postdoctoral Researchers

The available data indicate that more than half and as many as two-thirds of all postdoctoral researchers working in the United States are on temporary visas, and this percentage has been increasing consistently for decades.[14] According to GSS data, the percentage of temporary residents in the postdoctoral population has increased about 10 percentage points in the past 25 years. It has consistently been more than 50 percent in the physical sciences and 60 percent in engineering, and a surge of temporary residents in the life sciences—from about 30 percent to 50 percent—has accounted for almost all of the growth.

Independent research has found that among people earning a Ph.D. in the United States, noncitizens are more likely to pursue postdoctoral training (see Figure 2-6), and that noncitizen postdoctoral researchers produce more publications on average than do their citizen counterparts (Corley and Sabharwal 2007; Cantwell and Lee 2010; Stephan and Ma 2005). It is evident that the U.S.

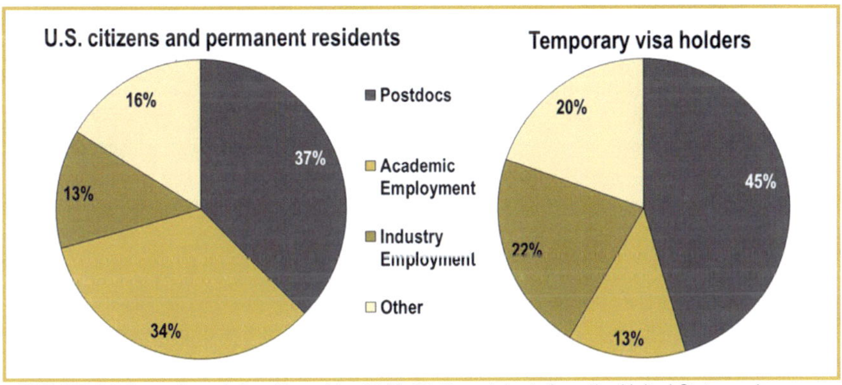

FIGURE 2-6 Commitments by citizenship, 2012. Ph.D. recipients from the United States only.
NOTE: "Other" includes government, nonprofit, other/unknown, and unreported sector but reported commitment to employment.
SOURCE: Data compiled from the NSF, NIH, U.S. Department of Education, U.S. Department of Agriculture, National Endowment for the Humanities, and NASA Survey of Earned Doctorates (SED): 2012.

[13] Although some institutions and disciplinary or professional societies have collected data on international postdocs in selected fields, there are no consistent and comprehensive data. This situation may be ameliorated with recent requirements for NIH grants to enumerate all laboratory personnel regardless of funding source.

[14] These include foreign nationals who received their Ph.D.'s in the United States as well as foreign-trained Ph.D.'s who first come to the United States for their postdoctoral position.

research enterprise is heavily dependent on noncitizen researchers, and therefore it would be prudent to know more about how many of such researchers there are, what their goals and expectations are, how satisfied they are with their training, and what should be done to make their experience more satisfying. The current system is dependent on a ready supply of highly skilled researchers who are willing to spend several years away from their home country, with only a moderate prospect for a permanent, full-time research position in this country at the end.[15]

Other Demographic Trends

The percentage of women in the postdoctoral population has grown between 10 and 20 percentage points in the past 25 years, depending on the field, with the life sciences increasing from below 25 percent to nearly 45 percent and the physical sciences and engineering increasing from below 10 percent to around 20 percent (GSS data). The percentage of women is higher among those who earned their Ph.D.'s in the United States than among those with foreign degrees. The rate of increase has been roughly the same in both populations.

The SED survey of those with U.S. Ph.D.'s indicates that the percentage of those with definite commitments for postdoctoral positions who are white is declining slightly, but whites still account for about 80 percent of the total. The GSS survey has not asked for ethnic information until the most recent survey years, so nothing definitive can be said about trends in those data. However, it is clear that some ethnic groups, particularly African Americans and Hispanics, are not proportionately represented in the postdoctoral population, either when compared to the U.S. population or with their representation in the population of doctoral degree recipients. The U.S. Census estimates underrepresented minorities (URM), including American Indian or Alaska Native, Black, Non-Hispanic, Hispanic, and Native Hawaiian and Other Pacific Islander, made up over 30 percent of the total population in 2012.[16] According to the GSS, URM U.S.-degreed postdoctoral researchers made up only approximately 9.5 percent of the total postdoctoral researcher population in 2012.[17] NIH director Francis Collins, among others, has acknowledged the problem and pledged to do more to increase diversity in the research community (Tabak and Collins 2011). None of the surveys ask about the ethnicity or country of origin of postdoctoral researchers who are temporary residents, even though they may make up over half of the total postdoctoral population. Studies of graduate students reveal that the largest sources of such students on temporary visas are China and India

[15] Additional information about the unique challenges of international postdoctoral researchers in the United States can be found in Appendix A.

[16] More information can be found at http://factfinder2.census.gov/faces/nav/jsf/pages/index.xhtml last accessed September 9, 2014.

[17] See also Einaudi 2013.

(CRS 2012), and it is likely that the composition of the postdoctoral population is roughly similar.

CONTEXT

Although postdoctoral researchers are developing new skills, a postdoctoral research position is also a "job." Practical matters, such as salary and benefits, career guidance, and personal concerns, are significant. The inherent short-term nature of postdoctoral employment can add complications. Moreover, immigration rules, sub-optimal communication skills, and cultural differences may pose a special challenge for postdoctoral researchers who are non-U.S. citizens.

Changes in the science and engineering enterprise—fueled by global collaborations, recent economic turmoil, and tight labor markets—have led to additional imbalances in the system. The postdoctoral population is growing more rapidly than the number of tenure track faculty positions at research universities. Among the results of this imbalance are that many individuals are staying as postdoctoral researchers for prolonged periods, and sometimes moving into a second or a third postdoctoral position; some attain research positions for which postdoctoral training was not necessary; and others are leaving their chosen research field or abandoning their research pursuits altogether.

In summary, research training beyond the Ph.D. continues to be a vital period for many researchers. However, the ideal experience defined a for postdoctoral researcher—"a temporary and defined period of mentored advanced training to enhance the professional skills and research independence needed to pursue his or her chosen career path"—is often not mirrored in reality. The characteristics of current postdoctoral experiences depend greatly on the source of funding, the research discipline, the approach of the principal investigator, and institutional setting.

3
Changing Aspects of the Postdoctoral Experience

The subtitle of the Committee on Science, Engineering, and Public Policy (COSEPUP) 2000 report is *A Guide for Postdoctoral Scholars, Advisers, Institutions, Funding Organizations, and Disciplinary Societies*, which reflects the 2000 Postdoctoral Report's intention to raise the visibility and official recognition of postdoctoral researchers among all stakeholders responsible for the quality and effectiveness of postdoctoral training.

There have been many changes in the past 14 years. Postdoctoral researchers themselves created the National Postdoctoral Association (NPA), as well as local postdoctoral organizations at numerous institutions. Many universities established offices to provide services, institute and communicate policies, and improve benefits. The National Science Foundation (NSF) assigned responsibility for overseeing postdoctoral training to the Division of Graduate Education (DGE) in their Education and Human Resources (EHR) directorate. The National Institutes of Health (NIH) launched a multifaceted program to meet the needs of its intramural postdoctoral researchers. Foundations supported efforts to survey postdoctoral researchers about their view of the system and to provide career guidance. Many professional and disciplinary societies also introduced programs aimed at helping postdoctoral researchers.

ACTIONS INSPIRED BY THE 2000 POSTDOCTORAL REPORT

Many of the actions to enhance the postdoctoral experience were consistent with the recommendations in COSEPUP's 2000 Postdoctoral Report, which called for improvements in data collection, policies and standards, and professional development (see Box 3-1 for details). A review of developments since 2000 reveals progress, but there remains a continuing need for further action in all these areas.

Data Collection

In 2002, the NSF Division of Science Resource Statistics (now called the National Center for Science and Engineering Statistics, NCSES) and the Commission on Professionals in Science and Technology (CPST) held a workshop titled "Postdocs: What We Know and What We Would Like to

Box 3-1
Summary of Recommendations from
***Enhancing the Postdoctoral Experience for Scientists and Engineers* (2000)**

- Data Collection
  - Collect comprehensive data on demographics, working conditions, career prospects
- Policies and Standards
  - Establish consistent policies across institutions and fields
  - Set a maximum total time limit for position
  - Include postdocs in policy planning
- Professional Development
  - Set standard/minimum salary scale, institutional status
  - Set standard/minimum benefits and access to institutional services
  - Conduct formal performance evaluations annually
  - Develop mechanisms for communications
  - Provide career guidance and skills training
  - Facilitate transition to career positions

SOURCES: National Research Council. "Enhancing the Postdoctoral Experience for Scientists and Engineers: A Guide for Postdoctoral Scholars, Advisers, Institutions, Funding Organizations, and Disciplinary Societies," Washington, DC: National Academy Press, 2000.

Know" (CPST 2002). Workshop participants included representatives from several leading professional societies (chemistry, math, biology, physics, and sociology), as well as NIH, the National Academy of Sciences (NAS) through COSEPUP, the American Association for the Advancement of Science (AAAS), Sigma Xi, and the NPA. The list of items that emerged in response to discussions included information about how postdoctoral positions fit into the career paths of researchers, what resources are available to postdoctoral researchers at individual institutions and at the national level, and the need to combine NSF data with that from professional societies. NSF promised to expand its data collection efforts, and in addition to some fairly major counting reforms to the GSS survey between 2007 and 2010 (Einaudi 2013), it did complete some additional surveying that was reported in the 2008 indicators. However, the fundamental limitations of the major surveys still exist.

When the NIH Biomedical Research Workforce Working Group (BMW Working Group) tried in 2012 to collect data on the number of postdoctoral researchers in the United States, "it quickly became clear that there are very little reliable data on the number of postdoctoral researchers.…This is due to a dearth of information about the numbers of foreign-trained postdoctoral researchers, as well as changes in the titles of postdoctoral researchers as they proceed through their training." The BMW Working Group made the following specific recommendations for data collection:

- The Survey of Earned Doctorates (SED) and Survey of Doctorate Recipients (SDR) should be modified to collect information about the career aspirations of doctorates and postdoctoral researchers.
- NIH should work closely with the NSF to shorten the lag between the collection and release of data.
- The NIH should work closely with NSF to ensure that the SDR includes those with foreign doctorates.

Two new surveys focusing on postdoctoral researchers are being developed at the NSF's NCSES: Survey of Postdocs at Federally Funded Research and Development Centers and Early Career Doctorates Project. The first of these has already produced one round of data, released April 24, 2013, with general information about the demographics, research fields, and support of the postdoctoral researchers working at federally funded research and development centers (FFRDCs) in the fall of 2010.[18] The second survey is a more ambitious project aimed to gain information about doctoral recipients of the past 10 years. Importantly, this survey will not be limited to either academic institutions, like the Survey of Graduate Students and Postdoctorates in Science and Engineering (GSS), or U.S.-degreed individuals, like the SED and SDR. A methodological survey was completed in early 2013, and plans appear to be underway to launch the full-scale project in 2014.[19]

Although the federal government has been the primary source of comprehensive data on the entire postdoctoral population, several other groups have used surveys to try to attain a clearer picture of postdoctoral working conditions, attitudes, aspirations, and priorities. With funding from the Sloan Foundation, Sigma Xi surveyed more than 7,600 postdoctoral researchers at 46 institutions (including intramural postdoctoral researchers on the NIH campus),[20] and collected a wide array of demographic data, information about family status, salary and benefits, training and education, and other administrative factors in postdoctoral researcher success (Davis 2005, Sigma Xi 2006). The Sigma Xi researchers found strong correlations between what they termed "administrative oversight and structure" and several measures of postdoctoral researcher success, including self-reported job satisfaction, ratings of advisers, low levels of conflict, and higher productivity as measured by

---

[18] More information can be found at http://www.nsf.gov/statistics/ffrdcpd/, last accessed April 4, 2014.

[19] More information can be found at http://www.nsf.gov/statistics/srvyecd/#sd&tabs-1, last accessed April 4, 2014.

[20] From the Summary of Survey Methods: "At each institution, Sigma Xi asked their contacts to assist in recruiting postdoctoral researchers who met the postdoc criteria set forth jointly by Sigma Xi and the institution in question. The researchers were either part of an institutional list, provided by local organizers, or were asked by the local organizer to register on a website so that Sigma Xi could eventually contact each one to ask the person to participate in the survey." More information is available at http://postdoc.sigmaxi.org/results/methods/. Last accessed on July 30, 2014.

number of publications. Interestingly, salary and benefits had a lower correlation with success than did other aspects of postdoctoral training. These findings are in line with other studies of knowledge workers in corporate environments, where nonmonetary factors, such as professional recognition and independence, often outweigh higher salaries in terms of job satisfaction (e.g., Herzberg 2003, Jayasingam 2013), although there are exceptions such as the IT industry (Thatcher et al. 2006).

More recently, the NIH Office of Intramural Training and Education (OITE) conducted a survey of its intramural postdoctoral researchers, with 43 percent of the more than 4,200 eligible postdoctoral researchers responding (NIH 2011). This survey focused on the mentoring of postdoctoral researchers with questions relating to setting research goals and expectations, opportunities for presenting and publishing research, and career guidance provided by mentors. The results of the 2010 study were also compared with a similar NIH study from 2001 and found small improvements in mentor availability, discussion of goals, and frequency of evaluations over the past decade.

Policies and Standards

NSF and NIH have taken steps to raise the profile of postdoctoral researchers and to explore ways to make postdoctoral training more effective. The NSF's DGE provides information on postdoctoral researcher-specific funding opportunities. The NIH Office of Postdoctoral Services, which is part of the NIH OITE, serves as a clearinghouse for information and policies affecting intramural postdoctoral researchers at the NIH and its affiliated institutes and centers. In 2009, they published the *NIH Postdoc Handbook*, which provides a comprehensive guide to policies and procedures of the complex NIH system.

In 2004, the Association of American Universities (AAU) surveyed university administrators representing 39 of its 62 member institutions. In addition to collecting basic demographic information about their postdoctoral researchers, the AAU survey included questions about salaries and benefits, administrative structures and policies, level of satisfaction about the postdoctoral experience, and suggestions for improvement (AAU 2005).

A large majority of the responding institutions reported that they stipulated minimum salaries (67 percent); provided employee benefits such as health insurance, vacation, and sick leave (87 percent); and maintained an office or administrative position to manage postdoctoral affairs (56 percent). What was most telling, however, is that only 10 percent conducted a postdoctoral researcher satisfaction exit survey, and a mere 8 percent collected data on postdoctoral researcher placement or career outcomes. Yet, 69 percent responded that postdoctoral training was working well at their institution.

In the late fall of 2011, the NPA conducted an informal survey of 175 institutions, 74 of which responded in full (an additional 18 institutions provided incomplete responses) (NPA 2012). Nearly all (94 percent) were institutional members of the NPA, and therefore had "already demonstrated a propensity to support postdocs and to improve their experience." Whereas appointment letters and annual reviews were standard at 89 percent and 54 percent, respectively, of the responding institutions, exit interviews were required in only 32 percent. Orientations were held for newly hired postdoctoral researchers at 85 percent of 74 responding institutions and more than 88 percent of 84 responding institutions had a required or recommended minimum salary.

In addition to these studies, individual universities have conducted internal surveys of their own postdoctoral researchers. The 2006 Sigma Xi study cited surveys at more than a dozen universities, medical schools, and research institutions. Because these surveys are intended for use by the institutions themselves, difficulties arise when trying to combine the data into a single comprehensive view.

Professional Development

One of the 10 action points identified in the 2000 Postdoctoral Report was to provide career guidance for postdoctoral trainees, and several institutions have moved in this direction. For example, the Burroughs Wellcome Fund has created a series of development guides for scientists at all stages of their careers. Although initially intended for biomedical researchers, much of the advice can be applied to other fields. The topics in this series include managing career transitions, obtaining tenure, giving research talks, and managing laboratory personnel (e.g., BWF 2006). The Burroughs Wellcome Fund also held workshops in 2002 and 2005 in conjunction with the Howard Hughes Medical Institute (HHMI) to offer career advice specifically to postdoctoral researchers and new faculty.

Allocation of personal professional development time has been recommended, most notably in the influential 2002 "SET for Success" report about science training in the United Kingdom by the Welsh physicist and advisor to the U.K. government Sir Gareth Roberts (commonly called the "Robert's report") . The report stated, "The Review believes that enabling the individual to establish a clear career path and a development plan to take them along it are critical to improving the attractiveness of postdoctoral research. The Review therefore recommends that [Higher Education Institutions] take responsibility for ensuring that all their contract researchers have a clear career development plan and have access to appropriate training opportunities—for example, of at least two weeks per year." It went on to recommend that funding agencies should make receiving research grants contingent on implementing such a standard (Roberts 2002).

Online tools for self-evaluation and career development, such as individual development plans (IDPs), enable postdoctoral researchers to organize their own career planning, set goals, identify needs, and keep track of progress (see Box 3-2). According to the NPA survey from 2011, around two-thirds of the responding institutions encouraged or required individual development plans, depending on the situation.

There are several online job search sites targeted specifically at postdoctoral researchers and many more that include postdoctoral positions in their listings.

Box 3-2
***MyIDP*: An Online Career Development Tool**

Individual development plans (IDPs) are often used to determine the best path for an employee by examining strengths, weaknesses, and goals, and how they fit with those of the employer. Although common in industry, this type of annual evaluation is not standard practice for many early-career scientists. However, publicly accessible resources are beginning to become available regardless of career development information provided by academic and research institutions.

An example is an online tool called *myIDP*: a web-based, personalized IDP designed for, but not exclusive to, postdoctoral scientists (students entering graduate school or transitioning to the workforce at a master's level would also benefit). Based on a program that was conceptualized by the Federation of American Societies for Experimental Biology (FASEB) in 2003, the current iteration of *myIDP* was developed by the American Association for the Advancement of Science, and authored by individuals from FASEB, the University of California, San Francisco, the University of Massachusetts Medical School, and the Medical College of Wisconsin. It is a four-step program that involves self-evaluation and assessment, career choice exploration, goal setting strategies and tools, and offline networking with mentors.

The site helps align (subjective, self-evaluated) skills, interests, and values of early-career scientists with potential careers within and outside of academe, and provides ample articles and resources to explore. In many cases, the information in these resources is not knowledge commonly held by their academic mentors and peers. It then allows the individual to set personal goals and time lines; it will also set up e-mail reminders to assist with keeping pace. The program encourages active mentor participation, although it acknowledges that academic mentors may not be suitable for some of the career-based goals outside of academe.

SOURCES: More information about *myIDP* is available at http://myidp.sciencecareers.org/. Accessed April 30, 2014.

Some of these sites have additional career information gauged for postdoctoral researchers, although the offerings can vary. One of the first career websites devoted to postdoctoral researchers was the Postdoc Network, launched in 2000 by the AAAS. The Postdoc Network provided an online forum for postdoctoral researchers (via a listserv), as well as job search capabilities and articles relating to career issues. This service has now been folded into the general career information provided by AAAS's *Science Careers* publication and website. *Nature* magazine provides similar listings and career advice on *naturejobs*.com and provides several online forums for postdoctoral researchers on its NatureNetwork site. These and other career resources for postdoctoral researchers can be found in Box 3-3.

Box 3-3
Selection of Career Resources for Postdoctoral Researchers

- MyIDP
- *Science Careers*
- PhDs.org
- MinorityPostdoc.org
- Vitae in the United Kingdom (www.vitae.ac.uk)
- Postdoc Career Development Initiative in the Netherlands (www.pcdi.nl)

Professional societies have traditionally played an important role in career development. With a cross-cutting view of a particular field of study, professional societies can engage and unite many different individuals, from undergraduates to emeriti professors (see Box 3-4).

INSTITUTIONAL CHANGES IN THE POSTDOCTORAL ENTERPRISE

In addition to changes that directly correlate with the recommendations of the 2000 Postdoctoral Report, there have been a number of cultural and institutional changes in the postdoctoral enterprise.

Box 3-4
Professional Society Job Sites

In deciding which next step to take in a career, professional societies play a key role. The electronic resources assembled and distributed by two particular societies—American Geophysical Union (AGU) Career Center and the Institute of Electrical and Electronics Engineers (IEEE) Job Site—are great examples.

The AGU's career website has not only the job listings and online resume hosting and search functions one would expect but also many other resources for early-career scientists. Articles, videos, workshops, and webinars provide a variety of opportunities for young scientists to explore options for their career beyond searching for their next postdoctoral position.

The IEEE Job Site provides the standard job search and resume posting as well as other useful tools for job seekers. IEEE ResumeLab assists members in creating a resume, CV, or cover letter; has a skill-assessment tool to help job applicants know what experiences to highlight; and even allows for mock interviews with a portfolio of potential questions. IEEE MentorCentre is an online service that helps young people connect with experienced professionals who can provide individual guidance and information. IEEE-USA Salary Service provides job seekers with information about benefits and salary based on data collected from more than 17,000 of IEEE's U.S. members. There is also a specific section for those seeking internships and entry-level jobs.

SOURCES: More information about the AGU Career Center is available at http://sites.agu.org/careers/. Accessed May 8, 2014. More information about the IEEE Job Site is available at http://careers.ieee.org/. Accessed May 8, 2014.

Professional Organizations for Postdoctoral Researcher

One common fact of life for postdoctoral researchers is relative isolation from their peers, especially in comparison with the more communal environment that they experienced in graduate school. A notable response to this situation during the past decade has been the creation by postdoctoral researchers of postdoctoral associations (PDAs), both at the national and institutional levels. Additionally, most research institutions have created postdoctoral offices (PDOs) at the administrative level or at least have appointed specific staff members to address the needs and concerns of postdoctoral researchers (e.g., see Box 3-5). A number of university PDOs organize career development activities, and campus PDAs often organize information sessions for their members. The NIH has an ambitious program of career-related activities for its intramural postdoctoral researchers (see Box 3-6).

The most significant initiative from the postdoctoral researchers themselves was the establishment of the NPA. In 2002, a group of postdoctoral researchers who were attending a meeting of *Science*'s Nextwave Postdoc Network formed the beginnings of the NPA with funding from the Sloan Foundation. The organization has grown to include more than 2,500 individual members plus 190 institutional members (as of December 2013). The NPA holds annual meetings

Box 3-5
The University of North Carolina at Chapel Hill: Fostering Diversity

Many underrepresented racial and ethnic minorities face overwhelming obstacles throughout their careers: isolation, stereotype threat, unconscious bias, little academic or social support, and few, if any, faculty mentors and role models who look like them.

The University of North Carolina (UNC) at Chapel Hill, in tandem with the Office of Postdoctoral Affairs (OPA), has spearheaded a national movement to develop peer-led minority postdoctoral organizations aimed at providing support, professional development, networking, and mentoring opportunities to strengthen the science, technology, engineering, and mathematics diversity pipeline. The goal is to bring together underrepresented African American, Hispanic, and American Indian postdoctoral scholars across disciplines to build valuable professional networks, and promote and support diversity and inclusion.

Similarly, the Carolina Postdoctoral Program for Faculty Diversity is one of the oldest diversity fellowship programs in the nation and has been highly successful. More than 30 years old, the program has supported more than 150 scholars since inception; 131 scholars teaching or working in higher education and 46 UNC faculty hires through the program. Support and mentoring are provided to postdoctoral researchers by senior-level administrators, department chairs, former fellows, and the OPA staff.

SOURCES: Thompkins, Sibby Anderson, Vanessa González-Pérez, and Jennifer Cohen. "Changing the Culture of Science: Minority Postdoctoral Organizations Help lead the Way." *The POSTDOCket* vol. 11, iss. 3, Summer 2013.; More information about the Carolina Postdoctoral Program for Faculty Diversity is available at http://research.unc.edu/offices/vice-chancellor/programs-projects-services/data_res_vcred_postdoc/. Accessed May 8, 2014.

Box 3-6
NIH Office of Intramural Training and Education

Home to about 4,000 postdoctoral researchers, the National Institutes of Health (NIH) plays a significant role in setting standards for the postdoctoral experience in the biomedical sciences. As mentioned elsewhere in the report, the NIH postdoctoral researcher salary levels are often used as standards by other institutions when establishing guidelines. Other practices they put in place for their postdoctoral researchers are less universally adopted, but are deserving of attention.

Upon arriving at NIH, postdoctoral researchers are given a comprehensive orientation and a Postdoc Handbook. The handbook covers relevant information, including workplace standards, guides to professional development, career guidance opportunities both inside and outside of NIH, and even financial- and immigration-related matters. This information is all curated by the Office of Intramural Training and Education, which also has a website that thoroughly covers these topics. Having a well-staffed (29 employees) central office to manage postdoctoral researcher affairs and providing common experience (e.g., orientation, career development workshops aimed at postdoctoral researchers) helps the postdoctoral researchers feel valued and part of a broader research community.

SOURCES: National Institutes of Health. "NIH Postdoc Handbook." Bethesda, MD: Office of Intramural Training and Education, 2012

to discuss issues affecting postdoctoral researchers, surveys its members to collect information about their experiences and opinions, educates policy makers at the national level, and provides training materials and other support to facilitate the creation and strengthening of both PDAs and PDOs at individual research institutions (see Box 3-7).

Mentor Evaluation

Mentors and advisors play a critical role in the postdoctoral experience due to the level of control they have over the postdoctoral researcher's career. Programs to train, evaluate, and recognize better mentors have started becoming more common. According to NIH, mentoring postdoctoral researchers can be counted toward effort reported on a research grant. "Postdoctoral researcher mentoring plans" have been required for all NSF grants supporting postdoctoral researchers since 2009,[21] as stipulated by the 2007 America COMPETES Act.[22] However, NSF does not subsequently evaluate how effectively the mentoring plan was implemented. Professional organizations, with input from postdoctoral members, have developed frameworks and guidelines to be used for this purpose, and compacts between postdoctoral researchers and their mentors, like

---

[21] More information can be found at http://www.nsf.gov/pubs/policydocs/pappguide/sf13001/gpg_2.njsp#IIC2j, last accessed June 19, 2014.
[22] NSF: SEC. 7008 (ACA 2007).

Box 3-7
National Postdoctoral Association: Offering Career Guidance

The National Postdoctoral Association's (NPA's) "Six Core Competencies" is a good demonstration of making postdoctoral researchers aware of what they should expect from their postdoctoral experience that will contribute to a successful career. These six key skills "are meant to serve primarily as: (1) a basis for self-evaluation by postdoctoral scholars and (2) a basis for developing training opportunities that can be evaluated by mentors, institutions, and other advisors." Along with the list, the NPA also gives details, examples, and additional resources for postdoctoral researchers to learn more about each of the competencies.

The Six Core Competencies are as follows:

1. Discipline-specific conceptual knowledge
2. Research skill development
3. Communication skills
4. Professionalism
5. Leadership and management skills
6. Responsible conduct of research

According to the NPA, "The goal of a postdoctoral fellowship is to provide the training necessary for the postdoctoral scholar to achieve intellectual and professional independence and success. This toolkit was developed to provide definition for the scholar and his/her support community as to the competencies essential to achieving this independence. Some of these competencies will have been acquired during graduate training. Thus, some aspects of the competencies will describe refinements of skills already achieved, and some will represent new or advanced skills."

SOURCES: More information about the NPA's Core Competencies is available at http://www.nationalpostdoc.org/competencies. Accessed May 8, 2014.

the one developed by the Association of American Medical Colleges (AAMC) Graduate Research, Education, and Training (GREAT) group (see Box 3-8; AAMC 2006), are an increasingly popular tool.[23]

As discussed in Chapter 2, the most common postdoctoral position involves a postdoctoral researcher working for a principal investigator on a research grant. Although the principal investigator's primary responsibility is the production of good research, the postdoctoral researcher is supposed to be receiving training, and the principal investigator is therefore the primary mentor. All principal investigators should take this responsibility seriously and do as much as they can to provide guidance to the postdoctoral researcher. However,

---

[23] According to the NPA survey, 36.5 percent encourage or require postdoctoral researcher supervisors to use a compact between postdocs and their mentors. These sometimes take the form of Individual Development Plans; see Box 3-2.

Box 3-8
GREAT Group Compact Between Postdoctoral Appointees

The Association of American Medical Colleges Graduate Research, Education, and Training (GREAT) group provides a forum for the 135 U.S. and 17 Canadian medical schools to develop policies for Ph.D. and M.D.-Ph.D. programs. Beginning in 2002, the GREAT group established a Postdoctorate Committee, which became the Postdoctorate Leaders Section in 2008, to enhance the quality of postdoctoral programs, provide professional development for postdoctoral researchers, and create consistent policies for postdoctoral researchers across member institutions.

The GREAT group published its Compact Between Postdoctoral Appointees and Their Mentors in 2006. The purpose was to provide a "model document" for institutions or individuals who want to set professional standards for a postdoctoral research position. The compact "is intended to initiate discussions at the local and national levels about the postdoctoral appointee-mentor relationship and the commitments necessary for a high quality postdoctoral training experience." Among the suggested uses (or benefits) of the compact is as "a recruitment tool to signify programmatic commitment to postdoctoral appointees."

SOURCES: Association of American Medical Colleges. "Compact Between Postdoctoral Appointees and Their Mentors." Washington, DC, 2006.

even if a principal investigator is committed to ensuring that the postdoctoral researcher is well prepared for her or his next career step, the investigator might not have the range of experience necessary to provide useful guidance. For example, a principal investigator who has worked only in a U.S. research-intensive university might not be able to help a postdoctoral researcher understand what is involved in a career in teaching-intensive liberal arts colleges, industry, government, or other countries. In addition, the crucial interpersonal rapport might not be right. Institutions where postdoctoral researchers work therefore have a responsibility to offer postdoctoral researchers a number of options for acquiring effective mentoring and to communicate to the postdoctoral researchers that they need to take some initiative in seeking out the guidance they want. Examples of sources of mentors can be found in Box 3-9.

Postdoctoral researcher-specific funding

The type of funding that postdoctoral researchers receive can have a dramatic influence on the quality of their experience. The NIH instituted a new funding mechanism in 2006, called the Pathway to Independence Award Program (K99/R00), which supports the transition of postdoctoral researchers to independent principal investigators. The journal *Cell* has published a list of other such "superpostdoc" funding programs in the United States and Europe (see Table 3-1, von Bubnoff 2007). Additionally, there are several career development tracks for postdoctoral researchers outside of laboratory research, including the National Academies' Christine Mirzayan Science and Technology

Box 3-9
Mentor Evaluation and Excellence Awards

The success of the postdoctoral experience often depends on various key mentoring relationships. Programs like the Presidential Awards for Excellence in Science, Mathematics and Engineering Mentoring Program (PAESMEM) and MentorNet are examples of attempts to recognize this important part of scientific research and seek to identify and honor outstanding mentors in science, technology, engineering, and mathematics (STEM).

The PAESMEM award, established by the White House in 1996, recognizes individuals, programs, or institutions that display "outstanding mentoring efforts that enhance the participation and retention of individuals (including persons with disabilities, women, and minorities) who might not otherwise have considered or had access to STEM." These are predominantly in higher education, though some K–12 programs have also received awards. After receipt, PAESMEM awardees can connect on the PAESMEM.net website, a forum for discussion and the exchange of mentoring ideas and best practices. The award and the website are administered by the National Science Foundation (NSF).

A PAESMEM award recipient in 2001, MentorNet is an online platform designed to "provide all STEM students with access to high quality mentoring relationships, helping mentors and protégés connect across generational, gender, racial, cultural, and socio-economic boundaries, and to create a vibrant and sustainable mentoring community that facilitates continuous engagement of protégés and mentors in mentorships throughout their educations and careers." The NSF noted, when conferring the award, "During the 2000-2001 program year, two thousand students were matched with one thousand nine hundred and thirteen mentors representing six hundred and ninety companies; seventy affiliated colleges and universities participated." Importantly, mentors come from diverse backgrounds, including industry, government labs, and academe. This has proven especially beneficial to women and others underrepresented in STEM fields.

SOURCES: More information about the PAESMEM awards and PAESMEM,net is available at http://paesmem.net/. Accessed May 8, 2014. More information about MentorNet is available at http://www.mentornet.net/ Accessed May 8, 2014.

Policy Fellowships and the AAAS Science and Technology Policy Fellowships. There are other programs designed to assist career transitions into media, teaching, and other professions.

Benefits

Postdoctoral researchers commonly fall between employee categories, and with this lack of solid definition comes incomplete or inconsistent benefits. Although postdoctoral researchers are often supplied with some form of official letter of appointment or contract, results from the 2005 Sigma Xi survey indicated that they are unaware of potential services available at their institutions or if those services are even offered (see Table 3-2).

TABLE 3-1 "Superpostdoc" Fellowship Programs in the United States and Europe

Program Country	Name
Germany	The Max Planck Society - Otto Hahn Prize
Austria	IMP Fellows Programme
United Kingdom	Sir Henry Wellcome Postdoctoral Fellowships
United Kingdom	Cambridge St. John's College Research Fellowships
United States	Harvard Junior Fellowship
United States	Carnegie Institution Staff Associate Program
United States	Whitehead Fellows Program
United States	Bauer Fellows Program at Harvard
United States	Rowland Junior Fellows Program at Harvard
United States	Lewis-Sigler Fellows at Princeton
United States	UCSF Sandler Fellows Program
United States	Sara and Frank McKnight Fellowships in Biomedical Research at UT Southwestern Medical Center
United States	CalTech Broad Fellows Program in Brain Circuitry
United States	Janelia Farm Junior Fellows

SOURCE: Updated from von Bubnoff 2007.

TABLE 3-2 Survey results from Sigma Xi survey (percent)

	Yes	No	Don't know
When you began this postdoc, did you receive an official letter of appointment or contract? [a]	84.3	12.1	3.0
Are you aware of formal written policies in your department or institution that address any of the following issues?			
• Determining paper authorship and author precedence [b]	21.4	49.4	29.2
• Defining misconduct [c]	38.7	34.2	27.2
• Resolving grievances [d]	28.0	39.1	32.9
• Determining ownership of intellectual property [e]	33.2	36.0	30.7
Are job placement services available to you at your institution? [f]	24.0	17.4	58.2
Is career counseling available to you at your organization? [g]	32.9	14.8	52.0

[a] 4,377 respondents; [b] 7,177 respondents; [c] 7,144 respondents; [d] 7,126 respondents; [e] 7,135 respondents; [f] 3,940 respondents; [g] 3,938 respondents.

SOURCE: Davis 2005

For the past 10 years, *The Scientist* has published an annual survey on the "Best Places to Work for Postdocs," which can provide insight into how a self-selected population of postdoctoral researchers views research institutions around the world (*Scientist* 2013). The magazine's survey is based on volunteered responses from life-sciences postdoctoral researchers, predominantly at U.S. institutions. Because the survey is not scientific, quantitative conclusions cannot be given much weight, but certain trends are clear; for example, which institutions have the highest proportion of postdoctoral researchers satisfied with certain aspects of their experience. Some institutions, such as the J. David Gladstone Institutes, consistently receive high marks by the postdoctoral researchers working there. Examples of some of the practices used at the J. David Gladstone Institutes are highlighted in Box 3-10.

Box 3-10
Supportive Practices for Postdoctoral Researchers

J. David Gladstone Institutes demonstrates that providing formal recognition and support structures can improve the postdoctoral experience.

J. David Gladstone Institutes is a medical research institution affiliated with the University of California, San Francisco. It includes graduate training for Ph.D. and M.D.-Ph.D. students, and many opportunities for postdoctoral researchers. It has been consistently ranked highly by *The Scientist* in its "Best Places to Work" list, both generally as an academic institution and specifically for postdoctoral researchers. Its treatment of postdoctoral researchers can serve as one example of good practices.

One area where J. David Gladstone Institutes excel is in open, easy-to-access information. Alongside the website for graduate students is a site of postdoctoral researcher-relevant material. This gives postdoctoral researchers an obvious place to find information and makes them feel like an integral part of the institution. The website also makes explicit many aspects of the postdoctoral experience that are often ill-defined. For example, the institute publishes standard salaries and benefits so that prospective postdoctoral researchers know what to expect and what is typical for all postdoctoral positions.

Another exemplary practice at J. David Gladstone Institutes is that they structure their postdoctoral positions so that the postdoctoral researchers can achieve some advancement in their careers (e.g., yearly raises and possible promotion to research scientist) over the course of their position, over and above additional research experience. The mentoring aspect of the position is taken seriously, with standardized expectations set by the institution. There is an Office of Postdoctoral Affairs, so that postdoctoral researchers have access to guidance and assistance in addition to what they receive from their principal investigator. Finally, postdoctoral positions at J. David Gladstone Institutes begin with salaries significantly above the standard set by the National Institutes of Health, and postdoctoral researchers receive raises of about 5 percent per year of seniority.

SOURCES: More information about the postdoc programs at the J. David Gladstone Institutes is available at http://gladstoneinstitutes.org/connect/postdocs. Accessed May 8, 2014.

Standard pay and benefits for postdoctoral researchers does vary among different institutions, and doctorate holders' hopes and expectations also vary. In the Sigma Xi survey of postdoctoral researchers, the benefits that were most desired were retirement savings benefits, dental insurance, and child care. Health insurance coverage is very important and expected. Sigma Xi found that expectation largely met, with 97 percent of postdoctoral researchers reporting health insurance offered to them with their job, although the survey did not indicate what portion of the cost the postdoctoral researchers covered and what portion of the cost was covered by the employing institutions. With this finding, and the NIH recently increasing the institutional allowance for health benefits associated with its National Research Service Awards, it appears that access to health insurance is among the areas where the actual postdoctoral experience is largely meeting expectations. The more recent NPA survey found that more than 78 percent of postdoctoral researchers who receive health benefits pay less than 25 percent of the premium. In addition, approximately 39.5 percent of 76 responding institutions provide paid maternity leave as a benefit to postdoctoral researchers and approximately 30.8 percent of 78 responding institutions provide "[c]ontributions to a retirement plan (postdocs allowed to contribute)" as a benefit to postdoctoral researchers (this number increases to 47.4 percent if the postdoctoral researchers are classified as employees).

In the survey of life-science postdoctoral researchers conducted by *The Scientist*, pay and benefits actually rank very low among the concerns covered by the survey. Much more common concerns, even at the institutions that were ranked highly, were about equitable treatment of postdoctoral researchers and allowances for family and personal life. Based on these and other findings, having an established set of guidelines and procedures is an important component to a satisfying postdoctoral experience (see Box 3-11). The NPA survey found that the majority of responding institutions (61 percent of 84) have established and disseminate a process for postdoctoral researchers to address their grievances (an additional 20 percent have a process established).

INTERNATIONAL ASPECTS OF THE POSTDOCTORAL EXPERIENCE

A major change in the postdoctoral enterprise is internationalization. Along with the increase in international postdoctoral researchers in the U.S. research system, the globalization of the science and engineering enterprise as a whole has dramatically altered the landscape.

Although it might seem reasonable to assume that the United States will continue to be able to attract large numbers of high-quality postdoctoral researchers from across the globe, it is necessary to remember that a policy change in China or India, better opportunities for postdoctoral research in other countries, or shifts in the labor market for those with postdoctoral training could also have significant, if less immediate, effects. The future well-being of the U.S. research system depends on understanding the changing dynamics of the

Box 3-11
Establishing a Formal Grievance Procedure

Many postdoctoral researchers do not have—or feel that they do not have—a way to deal with problems in the workplace outside of approaching their principal investigator. While the committee is not recommending any specific unionization or grievance procedure, it does see a need for some type of institutional mechanism for addressing problems that might arise for postdoctoral researchers.

Postdoctoral researchers in the University of California (UC) system unionized in 2008, selecting the United Automobile Workers Local 5810 as their exclusive representative, and negotiated their first contract with the UC system in 2010. The union represents more than 6,000 postdoctoral researchers across the UC campuses, and negotiates minimum standards for salary and benefits.

The collective bargaining agreement also contains a formal system for addressing postdoctoral researcher grievances. Under the union contract, if a postdoctoral researcher has a grievance, then she or he has a way of registering the complaint outside of her or his supervisor or department. The union also has established that the "grievance procedure allows us the option of taking a dispute to a neutral third party arbitrator for resolution, rather than to the University." The union contract set certain standards for working conditions and workplace protections, as well as salary and benefits, and violations of the contract can be cause for grievances.

SOURCES: More information about the UAW Local 5810 is available at http://uaw5810.org/about-your-union/history/. Accessed May 8, 2014.

global movement of scientists and engineers. Some research indicates that the relative attractiveness of study in the United States is declining. Although the U.S. reputation for quality research and faculty remains high, postdoctoral researchers have a less favorable impression of the U.S. lifestyle, and the employee benefits for U.S. postdoctoral researchers are considered inferior to those available in many other countries. (Stephan, Franzoni, and Scellato 2012)

Researchers found that international competition to attract highly qualified postdoctoral researchers is increasing. For example, a U.K. official said that "international recruitment is absolutely critical for the stability of the [English university] system at present time" (Cantwell and Lee 2010). Developed countries, such as Germany and Japan, have traditionally made efforts to encourage their citizens to return home after receiving postsecondary training in the United States. A recent trend indicates that many developing countries—including China and India, which together constitute the largest source of international graduate students in the United States (CRS 2012)—are actively recruiting their citizens who have conducted postdoctoral research abroad to return home for permanent positions. Similarly to the United States, postdoctoral organizations have formed around the globe, and the resources available to postdoctoral researchers interested in exploring international opportunities are numerous (see Box 3-12).

Box 3-12
Activities of Postdoctoral Organizations Outside the United States

Canada:

Canadian Association of Postdoctoral Students published an extensive survey of the Canadian postdoctoral researcher population in 2013.[a]

Japan:

National Institute of Science and Technology Policy in Tokyo surveyed postdoctoral researchers from seven universities and one national lab in 2005.[b]

United Kingdom:

The Vitae organization provides career support for doctoral researchers and also collects data through its Careers-in-Research Online Survey; the U.K. Research Staff Association, which is part of Vitae, serves as a professional organization for doctoral researchers.[c]

European Union:

Eurodoc serves doctoral candidates, postdoctoral researchers, and junior researchers.[d]

International:

The World Association of Young Scientists provides a global forum for issues concerning postdoctoral researchers and other young researchers.[e]

SOURCES:

[a] Mitchell, Jeremy, and Valerie Walker. "The 2013 Canadian Postdoc Survey: Painting a Picture of Canadian Postdoctoral Scholars." The Canadian Association of Postdoctoral Scholars/L'Association Canadienne de Stagiaires Post-doctoraux (CAPS-ACSP) and Mitacs, 2013.

[b] Misu, Toshiyuki, Tomoko Shimomura, Yukiko Miura, Akira Horoiwa, and Kan Imai. "Survey on Postdoctoral Fellows and Research Assistants (FY2005)." 1st Policy-Oriented Research Group National Institute of Science and Technology Policy (NISTEP) in cooperation with Knowledge Infrastructure Policy Division Science and Technology Policy Bureau Ministry of Education, Culture, Sports, Science and Technology (MEXT). Japan. August 2006. Available at http://www.nistep.go.jp/achiev/ftx/eng/mat128e/idx128e.html. Accessed on May 8, 2014.

[c] Centre, Careers Research and Advisory. "Vitae Researcher Development Framework." Cambridge, UK, 2011.

[d] More information about Eurodoc is available at http://www.eurodoc.net/. Accessed May 8, 2014.

[e] More information about the World Association of Young Scientists is available at http://ways.org/. Accessed May 8, 2014.

RECENT REPORTS

The National Academies are not alone in acknowledging a need to reconfigure the state of postdoctoral training in the United States. Although research institutions are clearly paying much more attention to the needs of postdoctoral researchers than they were in 2000, a flurry of recent reports suggests that many of the problems that existed in 2000 have yet to be resolved.

The volume of reports, articles, and discussions about postdoctoral training has grown enormously in the past 10 years.[24] One can interpret this as an either encouraging sign that the research establishment is finally focusing its attention on improving the lot of postdoctoral researchers or a dismaying alarm that the stresses within the postdoctoral system have become so pronounced that they can no longer be ignored.

Biomedical Research Workforce Working Group Report

Francis Collins, the director of the NIH, created the BMW Working Group in 2011 to evaluate the state of U.S. postdoctoral training. Published in 2012, the results of the *Biomedical Research Workforce Working Group Report* (BMW Report) correlate with much of what is presented in this report.

The BMW Report recommended funding a higher percentage of postdoctoral researchers through training grants and fellowships; experimenting with new ways to diversify postdoctoral training, raising salaries, ensuring adequate benefits, and increasing funding for "skip-the-postdoc" options, which encourage direct hiring into research career paths after a Ph.D.; and requiring that all postdoctoral researchers have individual development plans.

PCAST Report

The President's Council of Advisors on Science and Technology (PCAST) also released a report, titled *Transformation and Opportunity: The Future of the U.S. Research Enterprise*, which ends with a call to restructure postdoctoral training.

PCAST warned of a serious mismatch in the biomedical research labor market: "It thus seems that a significant fraction of today's postdoctoral fellows in biomedical research are essentially in training for jobs that do not exist in academia or for jobs in industry or other sectors into which they could move sooner. They are, *de facto*, low-paid university research staff." PCAST added that, "for the physical sciences, the problem is smaller, but similar in kind" (PCAST 2012b).

Professional Societies' Reports

A survey by the Computing Research Association found that the number of Ph.D. computer scientists entering postdoctoral positions doubled between 2008 and 2011, at a time when the number of openings for tenure-track faculty positions was falling steeply (CRA 2011).

An American Chemical Society commission recommended that "a) institutions, departments, and faculty mentors take greater responsibility for ensuring that postdoctoral associates develop professionally, b) all funding

---

[24] A Scopus search indicates that nearly 60 percent of all the articles published since 1953 with the topical keywords of postdoc, postdoctoral, or postdoctorate have been published since 2005; more than 30 percent since 2010.

agencies require general mentoring plans of applicants seeking support for postdoctoral associates, c) funding agencies become more receptive to requests for support of more senior research associates who are regular employees of research institutions" (ACS 2013).

A number of other organizations, professional societies, and media outlets have also written about the problems of postdoctoral training. There seems little doubt that this corner of the research enterprise deserves increased attention.

ADAPTING TO REALITY

The path forward needs to begin with recognition of the appropriate roles that postdoctoral researchers perform in the United States as well as the global science and engineering enterprise. Essential to functional and productive research activities, postdoctoral researchers can no longer simply be stop-gaps in individuals' career paths or serve as a source of highly trained, cheap labor. The policies of institutions and funding agencies must reflect and respect the needs of an increasingly diverse and international workforce and provide postdoctoral researchers with commensurate support and benefits. Likewise, postdoctoral researchers must approach and understand the commitments and expectations of postdoctoral positions in today's market. It is crucial and beneficial to everyone that guided self-examination about careers and career choice are made by graduate students and postdoctoral researchers.

4
The Shifting Career Landscape

Professional opportunities and career paths for scientists and engineers have changed dramatically over the past few decades. However, this shift has, for the most part, not been fully acknowledged within academe or the policy-making arena. Despite a growing body of evidence to the contrary, the dominant perception among mentors and funders is that the standard scientific career path goal remains a tenure-track faculty position at a research-intensive academic institution (references in Sauermann and Roach 2012). Compounding the issue, faculty advisors can be ignorant of—or biased against—what have been pejoratively termed "alternative" career paths and therefore are frequently incapable of advising their postdoctoral researcher on such options.

Efforts are being made by postdoctoral offices at the National Institutes of Health (NIH) and many universities to reform graduate and postdoctoral training so that it is helpful to those who pursue career paths other than tenure-track academic research (Box 4-1). The National Postdoctoral Association maintains a database of university policies and practices, which includes career-related activities. In response to changes in the science and engineering labor market, more attention is being paid to career planning beginning earlier in the career trajectories of scientists and engineers. As mentioned in Chapter 3, the *myIDP* program and the numerous career-planning workshops conducted by individual universities are steps in the right direction, but universities need to provide more information about what their graduates do when they leave, and professional societies should provide more information about career trajectories in their disciplines. Universities should also be aware of the incongruity of making postdoctoral training more relevant for career paths that do not require or reward advanced research training, which suggests that transition to such career paths ideally must occur before the postdoctoral period commences. Therefore, the information required for trainees to make informed decisions about such careers must also be available before the commencement of postdoctoral training.

UNDERSTANDING THE CAREER DEVELOPMENT OPTIONS

Even with the promising developments of the past decade, many graduate students and postdoctoral researchers do not have sufficient information about possible career paths, job prospects, and earning potentials to make informed

Box 4-1
Career Exploration Opportunities

Many postdoctoral researchers acquired Ph.D.'s without acquiring much work experience outside of academic research. Two programs—the Graduate Student Internship for Career Exploration (GSICE) at the University of California, San Francisco (UCSF) and the National Institutes of Health's National Institute of General Medical Sciences (NIGMS) biotechnology graduate research grants—address this issue by providing or requiring nonacademic experiences for graduates students.

The GSICE at UCSF recognizes that their Ph.D. recipients must be prepared for a variety of careers. The key aim is to include broader training in the UCSF graduate programs. This program, open to all graduate students in the basic and biomedical sciences, gives students work experience and provides career development workshops for 3 months (one academic quarter), and can be done either during or immediately following graduate school. Students can apply to be placed in companies in a variety of sectors, including education, business, and law. By encouraging students' exploration of other career paths, UCSF is not only preparing their graduates for the likely diversity of careers they will end up in, but also encouraging their students to think more broadly about their work and where it could lead them.

The NIGMS awards fellowships to institutions training graduate students in various areas of biomedical research, in order to support graduate research and high-quality training for graduate students. Grants awarded in biotechnology require recipient institutions to "include a two- or three-month industrial internship [during graduate education], to give students a meaningful research experience in a biotechnology or pharmaceutical firm. This research experience may be fully integrated with the trainee's Ph.D. research, but it may also be used by the trainee to delve into new areas." The NIGMS programs are currently established at 20 institutions across the country, and demonstrate the leverage that funding agencies have to affect the experience of graduate students and postdoctoral researchers.

SOURCES: More information about GSICE is available at http://gsice.ucsf.edu/. Accessed May 8, 2014. More information about the NIGMS biotechnology program is available at http://www.nigms.nih.gov/Training/InstPredoc/pages/PredocDesc-Biotechnology.aspx. Accessed May 8, 2014.

decisions at the appropriate stages to select realistic career goals. In addition, the National Science Foundation (NSF), which collects data on scientists and engineers, does not collect any data on the career outcomes of postdoctoral researchers who earned their doctorates in another country. Because a large fraction of today's postdoctoral researchers received their doctorates outside the United States, essential information simply is not yet collected.

Tracking Postdoctoral Researchers

Even for those who earned their Ph.D.'s in the United States, data about employment are quite limited. The economist Paul Romer conducted a revealing study in the late 1990s in which he asked a research assistant to request information about graduate study at 10 leading departments of biology, physics, chemistry, mathematics, computer science, and electrical engineering, and at 10 leading law and business schools (Romer 2001). None of the science and engineering departments provided information about the salaries of their graduates, nor did they provide information when it was specifically requested again in a later inquiry. Seven of the 10 business schools and four of the 10 law schools included salary information in the application packet. One more business school and three more law schools provided information in response to a later request.

The situation has apparently improved very little in the following decade. A 2008 survey of the websites of 15 top graduate programs in electrical engineering, chemistry, and biomedical sciences produced disappointing results. Of the 45 programs surveyed about placement of their graduates, only 2 had specific information, and 4 had some general information. By contrast, visits to 15 top economics department websites showed that 7 provided annual lists of where students found jobs. In addition, it should be noted that the collected information was for applicants to graduate school and not for postdoctoral researchers. In general, there is a stronger institutional interest in graduate students than in postdoctoral researchers; information about placement of postdoctoral researchers is virtually nonexistent (Stephan 2012).

There are some programs that attempt to follow the career trajectories of postdoctoral researchers, but the populations that have been studied are still very small and the data are difficult to find (Box 4-2).

Box 4-2
Postdoctoral Researcher Exit Surveys

Several institutions have implemented an exit survey, either through a postdoctoral office or postdoctoral association. These surveys provide institutions with an opportunity to gather valuable feedback from postdoctoral trainees that can be used to develop new policies and training programs, evaluate satisfaction with existing policies and programs, identify problems and areas for improvement, enable data-driven policy decisions, and track the career pathways of departing postdoctoral researchers. These surveys also provide an opportunity for postdoctoral researchers to give constructive feedback to their institutions, and are used by the institutions to establish a database of postdoctoral researchers and an alumni community.

Although implementation of exit surveys varies widely, there are common characteristics. Surveys are often provided online, and results are typically reported anonymously before being used to inform institutional policies. Survey questions tend to fall into three categories: personal and demographic information, details of the postdoctoral experience, and information on future plans. In addition to tracking the career pathways of departing trainees, the questions about the postdoctoral experience afford the postdoctoral researcher an opportunity to rank the performance of the institution and offer other feedback that may contribute to the improvement of existing programs.

Examples of institutional exit surveys

- Brown University – http://www.brown.edu/about/administration/biomed/graduate-postdoctoral-studies/end-appointment-process
- Indiana University School of Medicine - http://www.surveymonkey.com/s.aspx?sm=Mxfp3s_2f7Bx0EukMOuwlvtQ_3d_3d
- Medical College of Wisconsin (Password protected) – http://www.mcw.edu/postdoc/exitsurvey.htm
- The Scripps Research Institute - https://www.surveymonkey.com/s/pdexitsurvey_v4
- Stanford University (Password protected) - http://unmc.edu/postdoced/exitsurvey.htm
- University of Kansas Medical Center - https://survey.kumc.edu/se.ashx?s=5A1E27D26557444B
- University of Chicago Biological Sciences Division - https://www.research.net/s/Postdoc_Exit_Survey-Contact_and_Job_Info
- University of Nebraska Medical Center - http://unmc.edu/postdoced/exitsurvey.htm
- University of North Carolina at Chapel Hill - https://apps.research.unc.edu//exit_survey/survey.cfm
- University of Tennessee Health Science Center - http://www.uthsc.edu/postdoc/pdfs/Postdoc-Exit-Survey.pdf
- Washington University in St. Louis Division of Biology and Biomedical Sciences - http://dbbs.wustl.edu/Postdocs/Postdoc%20Alumni/Pages/PostdocAlumni.aspx
- Yale University (Password protected) - http://postdocs.yale.edu/postdocs/leaving-yale

SOURCES: National Research Council. “Enhancing the Postdoctoral Experience for Scientists and Engineers: A Guide for Postdoctoral Scholars, Advisers, Institutions, Funding Organizations, and Disciplinary Societies.” Washington, DC: The National Academies Press, 2000. More information http://www.nationalpostdoc.org/recommendations. Accessed May 8, 2014. Exit Survey examples last accessed April 17, 2014.

Attractiveness of the Academic Career Pathway

Assertions by policy makers about the career trajectories of postdoctoral researchers stem from calculations based on limited data. The data that do exist provide an overall picture of the career outlook for postdoctoral researchers and some detailed information about particular disciplines. For example, the NIH Biomedical Research Workforce Working Group compiled a flow chart for the career trajectories of U.S.-trained biomedical Ph.D.'s for their 2012 report; it indicated that more than 65 percent enter postdoctoral positions (BMW 2012).[25] With caveats about the quality of the data, the NIH advisory committee estimated that about 23 percent of biomedical Ph.D.'s move into tenure-track faculty positions (see Table 4-1 for a complete breakdown).

Survey research by Sauermann and Roach found that postdoctoral researchers do have a relatively accurate impression of the percentage of Ph.D.'s who find tenure-track research faculty positions within five years (see Table 4-2, Sauermann and Roach 2013), but this apparently does not discourage a large percentage of graduates from pursuing postdoctoral training for that purpose.[26] Sauermann speculates that, because they have always been at the top of the class and work at top-tier institutions, postdoctoral researchers assume that they also will be the lucky ones who obtain their desired jobs. Such findings highlight the fact that before becoming postdoctoral researchers, graduate students need more disaggregated information to truly understand the labor market. They especially need to know the outcomes for postdoctoral researchers at specific institutions.

As was highlighted in the BMW Report, a significant number of graduates pursue other types of research careers, including many that are open to people without postdoctoral experience. Nearly one-third enter positions that do not use

TABLE 4-1 Career outcomes for biomedical Ph.D. recipients

Position	Percentage
Tenure-Track Faculty Positions	23
Non-Tenured Academic Research or Teaching	20
Industrial Research	18
Government Research	6
Science-Related Positions (Non-Research)	18
Non-Science-Related Positions	13
Unemployed	2

SOURCE: BMW 2012

[25] During the writing of the Biomedical Workforce Report, the rate at which U.S.-trained doctorates with definite commitments entered into postdoctoral positions for the life sciences was closer to 70 percent.

[26] The Sauermann and Roach survey was conducted via on-line web survey at 39 tier-one U.S. research universities with doctoral programs in science and engineering fields determined by consulting the National Science Foundation's reports on earned doctorates (Materials and Methods section of Sauermann and Roach 2012).

TABLE 4-2 Results of the Sauermann and Roach Postdoctoral Researcher Survey (percent)

	Life Sciences	Chem-istry	Physics	Engin-eering
Ph.D.'s holding a tenure-track faculty position 5 years after graduation				
• S&E* Indicators, 2012	14.3	16.5	16.5	15.5
• Postdoc' Estimate	14.7	15.5	13.3	17.1
Pursued a postdoc primarily to obtain a tenure-track faculty position in the future	61	55	63	47
• If yes, probability that they will hold a tenure-track faculty position in 5 years	55	68	42	58
Desired position for Ph.D.'s				
• Faculty-Teaching	23	20	22	9
• Faculty-Research	30	14	29	25
• Government	10	13	14	14
• Established Firm	18	34	17	29
• Start-up Firm	9	10	13	19
• Other	10	9	5	5
Desired position for Postdocs				
• Faculty-Teaching	18	15	15	9
• Faculty-Research	44	41	50	44
• Government	14	19	19	18
• Established Firm	13	15	6	21
• Start-up Firm	5	7	5	7
• Other	6	3	5	2

*S&E: Science and Engineering

SOURCES: Sauermann and Roach 2013, and unpublished presentation by Sauremann to the Committee on Science Engineering and Public Policy on September 9, 2013, available at http://sites.nationalacademies.org/PGA/cs/groups/pgasite/documents/webpage/pga_084875.pdf, accessed on July 24, 2014.

their research skills (BMW 2012, Kahn 2011). Moreover, the data used to determine this level of skill mismatch are now "dated" and reflect the period when the NIH budget was doubling. In the flat funding environment of today, these data likely overstate the possibility of attaining a research position. Although independent postdoctoral research training could conceivably have some value for someone in a non-research career, in many, if not most cases, the time spent in a postdoctoral position might have been more productively spent acquiring experience directly relevant to that career (BMW 2012).

Although the federal government has assumed most of the responsibility for collecting data about postdoctoral researchers, the institutions where they work are actually in the best position to track what postdoctoral researchers do after completing their training. The current effort to match the UMETRICS data for graduate students and postdoctoral trainees supported on federal grants at 12 universities to U.S. Census data is one promising experiment (Weinberg 2014). Collecting and making available information on the subsequent career paths of postdoctoral researchers would be very helpful to Ph.D.'s deciding whether to pursue postdoctoral training or not and to the nation's research agencies, which must decide if their investment in postdoctoral training is producing the desired results (see Box 4-3).

Box 4-3
The United Kingdom's Concordat to Support the Career Development of Researchers

A common complaint among postdoctoral researchers is the ambiguity of the position. Expectations and experiences vary among institutions and among principal investigators, making it difficult to determine if one is receiving proper treatment and management. In an effort to clarify what postdoctoral training should entail, a coalition of institutions in the United Kingdom, managed by Research Councils UK, developed the Concordat to Support the Career Development of Researchers, which has been endorsed by funding agencies, prominent scientific institutions, major universities, and professional societies.

The Concordat establishes important principles and sets explicit standards for the roles of principal investigators, mentors, supervisors, and the postdoctoral researchers themselves. The Concordat sets out seven key principles for funders and employers, including recognition of researchers' role in their host institution and the importance of personal and career development. The principles include a system of oversight, with one principle being that "[t]he sector and all stakeholders will undertake regular and collective review of their progress in strengthening the attractiveness and sustainability of research careers in the UK." For example, under the Concordat, regular surveys are done to track whether scientists are given the opportunity (and take the opportunity) to participate in management training. Additionally, many awards take into consideration institutional compliance with the principles in the Concordat.

SOURCE: Vitae. "Concordat to Support the Career Development of Researchers: An Agreement between the Funders and Employers of Researchers in the UK" https://www.vitae.ac.uk/policy/vitae-concordat-vitae-2011.pdf. Last accessed May 8, 2014

POSTDOCTORAL RESEARCHERS SUPPLY AND USE

Policy recommendations from the research community repeatedly warn of imminent shortages of graduates in science and engineering,[27] but some job market data do not indicate a shortage. In many fields, more doctoral recipients are being produced in the United States and abroad than the market for research positions demands, given current levels of funding. (Stephan 2012). In addition, most of these studies focus on undergraduate, masters, and professional degree programs, not postdoctoral researchers.

The supply of domestic doctorates depends on the number of Ph.D.'s who have recently been produced, and the options they have for other types of employment. In recent years, the supply, especially in engineering and the biomedical sciences, has been growing, while job opportunities outside of academia have not matched this increase. As a result, fewer new Ph.D.'s have definite commitments at the time they graduate, and those who do have plans are more likely to take a postdoctoral position (Figure 4-1).

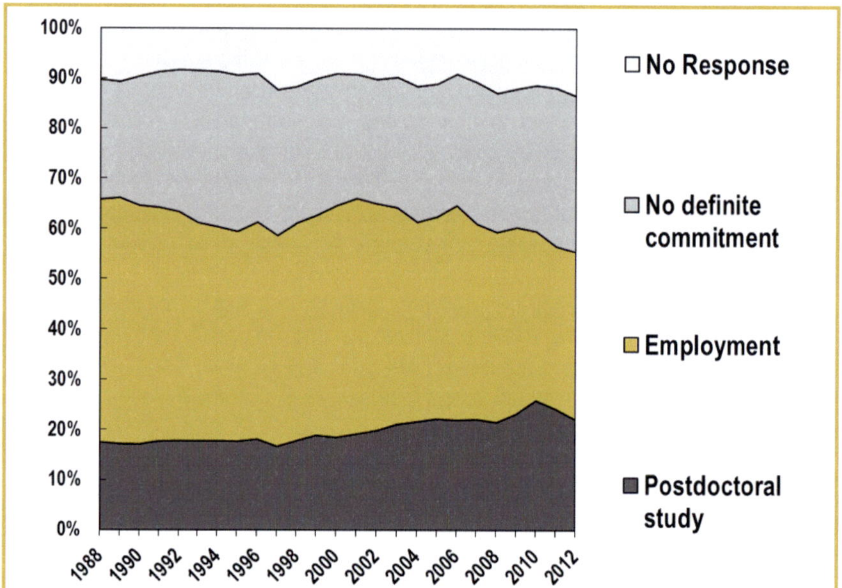

FIGURE 4-1 Postgraduation plans of U.S.-trained doctorate recipients.
SOURCE: Data compiled from the NSF, NIH, U.S. Department of Education, U.S. Department of Agriculture, National Endowment for the Humanities, and NASA Survey of Earned Doctorates: 2012, 2011, 2010, 2009, and 2008.

[27] For example PCAST 2012a.

In addition, the number of postdoctoral researchers per faculty position has more than doubled since the early 1970s.[28] Each year, U.S. universities produce almost 28,000 new Ph.D.'s in science and engineering alone (GSS data). Moreover, a growing number of foreign-trained doctorates, notably from emerging economies such as China and India, add to supply. These trends have shown a dramatic increase in the past decade, especially in engineering and the biomedical sciences.[29] At the same time, the percentage of doctoral recipients obtaining tenure-track faculty positions has been declining, and this trend does not show any sign of reversal (Table 4-3). Therefore, it would appear that far more doctoral recipients are pursuing postdoctoral training than there are job openings that require such training.

In other words, for many the postdoctoral researcher position is becoming the default after the attainment of the Ph.D., especially in the life sciences, and the decision to seek a postdoctoral position is often made without regard to whether advanced training in research is really warranted for the individual in question.

SALARY

A major point of discussion is the historically low salary paid to postdoctoral researchers. The demand for workers with certain types of training and the salaries offered for different types of expertise are affected by a number of factors such as the business cycle, government research funding, industrial research and development budgets, and immigration policy. As with actual markets, the lower the salary, the greater the demand will be from principal investigators to hire postdoctoral researchers if it is perceived that doing so provides greater research output than hiring either staff scientists or additional graduate students.[30] The research enterprise does not have a fixed need for a certain number of postdoctoral researchers; instead, principal investigators will typically hire a larger number of Ph.D.-trained workers when their funding allows.

---

[28] According to the NSF-NCSES special tabulations (2013) of the 1973–2010 Survey of Doctorate Recipients, the ratio of U.S. trained postdoctoral researchers to full-time faculty increased nearly 170 percent from 1973. In addition, according to the GSS, between 2000 and 2010, the number of postdoctoral researchers increased from 43,115 to 63,415, or a 47 percent increase and the number of graduate students increased from 493,311 to 626,820, or a 28 percent increase. Over that same time period, the Department of Education's Integrated Postsecondary Education Data System (IPEDS) Salaries, Tenure, and Fringe Benefits Survey for Faculty reported only a 13 percent increase in tenure and tenure track faculty positions (from 369,784 to 421,370), and a 21 percent increase in the total faculty (from 490,861 to 604,393), but a 45 percent increase in "Instructors, Lecturers, and Other" (non-tenure/tenure-track) faculty positions (from 121,077 to 183,023). The ratio of postdoctoral researchers to tenure and tenure-track faculty, according to GSS/IPEDS data, increased nearly 150 percent since 1975.

[29] More details on these trends may be found in Chapter 2.

[30] Discussed further below.

TABLE 4-3 Employed U.S.-trained science, engineering, and Health (SEH) doctorate recipients holding tenure and tenure-track appointments at academic institutions, by field of and years since degree: 1993–2010 (Percent)

Years since doctorate and field	1993	1995	1997	1999	2001	2003	2006	2008	2010
< 3 years									
All SEH fields	18.1	16.3	15.8	13.5	16.5	18.6	17.7	16.2	14.7
Life sciences [a]	9.0	8.5	9.3	7.7	8.6	7.8	7.2	6.5	7.6
Computer/information sciences	31.5	36.5	23.4	18.2	20.7	32.5	31.2	22.0	20.8
Mathematics and statistics	40.9	39.8	26.9	18.9	25.2	38.4	31.6	31.3	26.1
Physical sciences	8.8	6.9	8.5	7.8	10.0	13.3	9.8	8.8	6.8
Psychology	12.8	13.6	14.7	16.0	15.6	14.6	17.0	18.1	16.0
Social sciences	43.5	35.9	37.4	35.4	38.5	44.8	39.3	45.4	41.1
Engineering	15.0	11.5	9.4	6.4	11.3	10.8	12.4	9.3	7.5
Health	33.9	34.2	30.1	28.1	32.1	30.3	36.2	27.7	24.2
3–5 years									
All SEH fields	27.0	24.6	24.2	21.0	18.5	23.8	25.9	22.9	19.7
Life sciences [a]	17.3	17.0	18.1	16.4	14.3	15.5	13.7	14.3	10.6
Computer/information sciences	55.7	37.4	40.7	25.9	17.3	32.2	45.7	37.8	22.2
Mathematics and statistics	54.9	45.5	48.1	41.0	28.9	45.5	50.6	40.7	41.7
Physical sciences	18.8	15.5	14.5	11.9	15.8	18.3	19.7	16.5	14.7
Psychology	17.0	20.7	16.8	17.6	17.5	19.9	23.8	18.3	19.1
Social sciences	54.3	52.4	50.4	46.5	38.8	46.0	50.4	48.9	46.7
Engineering	22.7	19.3	19.4	12.6	10.8	15.9	16.3	15.5	13.0
Health	47.4	40.2	41.1	39.5	25.1	40.8	43.1	34.4	33.3

[a] Includes Biological, agricultural, and environmental life sciences.

NOTES: Proportions are calculated on the basis of all doctorates working in all sectors of the economy. Data for 1993–99, 2001, and 2006 include graduates from 12 months to 60 months prior to the survey reference date; data for 2003, 2008, and 2010 include graduates from 15 months to 60 months prior to the survey reference date.

SOURCE: National Science Foundation, National Center for Science and Engineering Statistics, Survey of Doctorate Recipients (1993–2010), http://sestat.nsf.gov. Science and Engineering Indicators 2014

Current Salary Levels

As described in Chapter 2, biomedical postdoctoral researchers are the largest segment of the postdoctoral population. As a result, NIH policies tend to set the tone for all postdoctoral positions in terms of salary, benefits, and duration.[31] In many institutions, and especially in the biomedical sciences, the NIH National Research Service Award (NRSA) postdoctoral salary rate becomes "the" salary that all postdoctoral researchers are paid.

The NIH NRSA stipend for beginning postdoctoral researchers in 2004 was $35,568, which would be $43,230 in 2012 dollars (or $44,207 in 2014 dollars). Despite repeated calls to raise postdoctoral salaries, the NRSA stipend was increased to only $39,264 in 2012. In 2014, NIH raised the stipend to $42,000, which, in real terms, is actually lower than the 2004 level. Institutions that do match the NRSA minimum tend to lag by 6 to 8 months, and many have yet to implement this most recent increase. Whereas some institutions do not match the NRSA stipend level, others pay considerably more. An extraordinary example is that of the postdoctoral researcher salary rates at the Los Alamos National Laboratory, where beginning postdoctoral researchers received a salary of $72,200 in 2013.[32] These institutions, however, also often use the postdoctoral position as a way of screening future employees, something that is rarely done in academe.

Principal investigators, and some funding agencies, have recognized the financial advantages of hiring postdoctoral researchers rather than graduate students or staff scientists to work on research grants. Graduate student stipends coupled with tuition charges are often on par with—or exceed—standard postdoctoral researcher salaries, especially at private universities.[33] Postdoctoral researchers also come with more experience than graduate students, do not require the same level of supervision and training, and according to the 2006 Survey of Doctorate Recipients, work between 2,500 and 2,650 hours a year depending on field (Stephan 2012), compared to just over 2100 hours a year for the average full-time worker the United States.[34] Staff scientists have research expertise comparable to postdoctoral researchers, but standard salaries and

---

[31] The NPA survey from late fall of 2011 found that at 47 percent of the responding institutions "the minimum salary or stipend was the NIH National Research Service Award (NRSA) stipend for 0 years of experience or gave the equivalent amount of $38,496. Other responses (41 percent) reported dollar amounts ranging from $28,000 to $55,000…, with most responses in the $35,000–$38,000 range." More information can be found at http://www.nationalpostdoc.org/images/stories/Documents/Other/npa-survey-report-april-2012.pdf, last accessed April 4, 2014.

[32] More information can be found at http://www.lanl.gov/careers/career-options/postdoctoral-research/postdoc-program/postdoc-salary-guidelines.php, last accessed April 9, 2014.

[33] Compiling data from GradPay—a crowdsourcing website set up to collect information about graduate student stipends—and IPEDS Institutional Characteristics Survey Tuition Data, it was determined that the average total cost for a science, engineering, or health graduate student in 2011 was approximately $50.7 thousand, ranging from around $40 thousand to an extreme of $94 thousand. Data were gathered for 73 institutions and 51 self-identified science, engineering, or health fields. More information about GradPay is available at http://gradpay.herokuapp.com/, last accessed April 8, 2014.

[34] More information can be found at http://www.bls.gov/cps/cpsaat19.pdf, last accessed September 9, 2014.

benefits for these positions can be twice as high as a typical postdoctoral stipend. Clearly, postdoctoral researchers can be a bargain compared with many graduate students and staff scientists, from the perspective of a principal investigator. In fact, postdoctoral salaries, which start on average at about $41,000 a year, are approximately 52 percent of the average nine-month salary for newly hired assistant professors at public research universities (OSU 2014) and have not kept pace with other positions, making postdoctoral researchers an even more financially appealing choice.[35]

The starting salary is also low compared with other employment sectors. New Ph.D.'s who do not take postdoctoral positions at universities, for example, can command a starting salary that is 40 to 200 percent more than their counterparts, depending on their field and sector of employment (Survey of Earned Doctorates data). The postdoctoral starting salary is low even relative to that of those who earned only a bachelor's degree who, by the time they reach their late 20's to early 30's, were earning about $49,911 in 2012.[36] If one assumes that this is actually a market economy, the question arises as to why postdoctoral salaries are so low. One part of the answer is that the presence of temporary residents in the labor market, coupled with the large supply created by Ph.D. programs in the United States and the lack of alternative types of positions, puts downward pressure on postdoctoral wages and makes it possible for principal investigators to fill their laboratories with newly minted Ph.D.'s at "rock-bottom prices." [37] Further, the number of postdoctoral researchers is largely set by the availability of research support (typically federal), rather than by the availability of suitable research positions after the postdoctoral period. Thus the scientific labor market for postdoctoral researchers does not function like a labor market in the usual economic sense of the term.

The Value of Training

The traditional rationale put forward to justify the low salaries paid to postdoctoral researchers is that advanced training positions will qualify the postdoctoral researcher for a better-paying position in the future. However, this argument can be challenged on several fronts.

First, there is no guarantee that the postdoctoral researcher will receive significant experience to prepare her or him for full-time independent research

---

[35] The average nine-month salary of a newly hired assistant professor at public research universities in the biological and biomedical sciences is $74,177; in engineering it is $84,012; in math and statistics it is $67,383 (OSU 2014).

[36] U.S. Census Bureau, Current Population Survey, Annual Social and Economic Supplements. Table P-28. "Educational Attainment--Workers 18 Years Old and Over by Mean Earnings, Age, and Sex: 1991 to 2012" More information can be found at http://www.census.gov/hhes/www/income/data/historical/people/, last accessed June 9, 2014.

[37] For many non-U.S. doctorates, a $41,000 salary and the possibility of ultimately becoming a U.S. citizen is a very appealing option. Thus, it is not surprising that a majority of U.S. postdocs are on temporary visas, and a majority of these postdocs earned their Ph.D.'s in other countries.

positions. Roughly 80 percent of postdoctoral researchers are employed by principal investigators and paid out of research grants. The principal investigators are currently under no obligation to provide opportunities for development, such as experience writing grants, managing a laboratory, or giving presentations. This means that, in many cases, any training that occurs is a byproduct of work, rather than planned career development activity.

Second, as the discussion above indicates, the number of job openings each year that require postdoctoral training is far lower than the number of people who acquire postdoctoral positions each year. Although no one can be promised a job just because he or she completes a postdoctoral appointment, those entering a training program should know how well their predecessors fared in the job market.

Third, to the extent that training occurs, postdoctoral researchers pay an extremely high price for it. As mentioned, postdoctoral researchers work approximately one-third more hours than the average full-time worker in the United States,[38] yet earn on average about 20 percent less than people of a comparable age who have only a bachelor's degree. That works out to approximately 40 percent less per hour.

Fourth, when postdoctoral researchers do eventually pursue nonacademic or non-research jobs, they do not receive a wage premium for their additional training. Overall, the sacrifices made by postdoctoral researchers in salary and benefits are not compensated later in their careers. On average, postdoctoral researchers start at lower salaries than what is paid to graduates who entered similar jobs immediately after earning their Ph.D., even though a postdoctoral researcher has had several years of additional research experience (Kahn 2011). Additionally, the total earnings of postdoctoral researchers trail those of their Ph.D.-only contemporaries for the rest of their careers (Kahn 2011). Employers appear to be sending a signal that the time spent in postdoctoral research is not valued in many job markets.

INFORMING DECISIONS

Doctorate recipients are adults who have the right to choose a career path and accept low wages if they think that it will eventually lead to a satisfying career. However, there is a difference between abstract knowledge and concrete understanding of the prospects of those individuals who attend the same institution they are considering. To make informed decisions, they need detailed information on the current job market and the job results for those who have completed postdoctoral training broken down by field and institution. Too few institutions are collecting and disseminating those data.

---

[38] More information can be found at http://www.bls.gov/cps/cpsaat19.pdf. Last accessed September 9, 2014.

5
Recommendations

Postdoctoral researchers are a significant, but often overlooked, segment of the science and engineering research workforce. Many different types of positions come under the postdoctoral researcher designation, but an appropriate umbrella term that describes these individuals is the current definition agreed upon by the National Institutes of Health (NIH), National Science Foundation (NSF), and the National Postdoctoral Association (NPA): "An individual who has received a doctoral degree (or equivalent) and is engaged in a temporary and defined period of mentored advanced training to enhance the professional skills and research independence needed to pursue his or her chosen career path." Although the individual postdoctoral experience varies significantly depending upon a number of factors such as location, field, or funding source, as examples, there is little debate about the potential value that the general postdoctoral experience provides to either the postdoctoral researcher or to his or her host institution.

Over the past 20 years, the percentage of new Ph.D.'s with definite commitments taking postdoctoral positions has increased in all fields, reaching a recent peak in 2010 when the American Recovery and Reinvestment Act provided a temporary boost in research funding. Among research disciplines, this growth has been most rapid in engineering and the social sciences—fields in which postdoctoral training was relatively uncommon a decade ago. Comparing the various sources of funding, research positions funded by a principal investigator's grant are the most common and have also seen the largest increases in the past decade. The demographics of the postdoctoral population have also been changing: there are more women and more temporary residents, and their median age has increased, as scientists are spending more time in postdoctoral positions.

Although the broad trends are known, exact statistics about the changing nature of postdoctoral positions and researchers have significant uncertainties. Information on the actual number of postdoctoral researchers and how they are supported is difficult to obtain and those data that do exist are often incomplete, covering only certain subsets of the postdoctoral population. In addition, most funding agencies and research institutions do not track the career outcomes of postdoctoral researchers.

The problem of incomplete data is linked to the problems with the postdoctoral experience itself. The paucity of data concerning the number and characteristics of postdoctoral researchers in the United States is due in part to their poorly defined status at many institutions, the wide variety of titles applied to postdoctoral researchers, and the number of postdoctoral researchers who come to the United States subsequent to receiving their doctoral training abroad. Unlike undergraduates, graduate students, staff, and faculty, which are well-organized groups, postdoctoral researchers are not a well-defined population at many institutions and therefore can be invisible to administrators.

Research practices and expectations of postdoctoral researchers are quite different across disciplines and institutional settings, and these variations are translated into differences in postdoctoral experiences. In general, the practice of employing postdoctoral researchers as long-term researchers, with little mentoring and little hope of moving into a career that requires advanced research training, is becoming more common. The mentored training aspect of a postdoctoral researcher's experience can be inconsistent and often inadequate. The mismatch between the expectations and outcomes of the postdoctoral experience causes disappointment and disillusionment for some postdoctoral researchers, and may discourage undergraduate students and graduate students from continuing to pursue careers in research, thereby reducing the pool of talent on which the research enterprise depends.

Although there have been a number of improvements since the release in 2000 of the National Academies' report *Enhancing the Postdoctoral Experience for Scientists and Engineers*, postdoctoral researchers at many institutions continue to lack adequate mentoring, recognition, status, and benefits. Many institutions do not have a coherent set of policies, practices, and procedures for postdoctoral researchers that are equivalent to those available for students, faculty, or staff, and many postdoctoral researchers do not know about those policies that do exist. This lack of support structure and official status is often cited as a bigger concern than salary issues in studies of current postdoctoral researchers.

In addition, there is a lack of data on the career aspirations, preferences, and reasons that influence graduate students and postdoctoral researchers to pursue research careers. It appears that many Ph.D. recipients have been conditioned to see a postdoctoral position as the logical next step in their career progression, without careful consideration as to whether advanced research training is required to further their career goals. Although it is ultimately the individual doctorate holder's decision, it is unclear whether they or their faculty mentors have sufficient resources to make a fully informed choice.

There is a continuous need for researchers with advanced training in the U.S. research enterprise. Postdoctoral researchers are playing a crucial, but often unrecognized, role in research. They are contributing significantly to academic research and they fill important roles in research groups at national laboratories, in government, and in industry. However, some principal investigators hire postdoctoral researchers to fill the need for advanced researchers in lieu of permanent research staff, instead of as a symbiotic practice that provides advanced training. Unfortunately, there is some evidence that this practice is increasing.

Given the current levels of total research spending in the United States, the practice of hiring postdoctoral researchers to staff laboratories has created a situation where the number of postdoctoral researchers is out of equilibrium with the number of available positions that require advanced training, and there is no reasonable correlation between the change in the total number of postdoctoral researchers and positions that require postdoctoral training. Significantly fewer than half of all postdoctoral researchers continue into academic tenure-track positions and an increasing fraction end up in nonacademic or non-research careers that do not require the years of advanced research training provided by the postdoctoral position.

Because of this mismatch, postdoctoral training does not always contribute to the career advancement of postdoctoral researchers. There is a need to reexamine the human capital needs (i.e., job structure, salary practices, and career pathways) of the research enterprise. Some of the work now being done by postdoctoral researchers might more appropriately be done by permanent research staff, who receive the salary, benefits, and job security commensurate with full-time employment. Such research staff positions are common in government, industrial laboratories, and outside the United States. The postdoctoral experience itself should be refocused, with training and mentoring at its center.

Graduate students should be made aware of the wide variety of career paths are open to them. For some careers, particularly for faculty positions in the physical and biomedical sciences at research universities, the postdoctoral experience can be very helpful. However, for many careers, a new Ph.D. can benefit more from other types of work experience—a postdoctoral position is not the only way to enhance one's skills and advance one's career.

The primary focus of this report is on the largest segment of the postdoctoral population: postdoctoral researchers working at universities and being paid as part of a principal investigator's research grant. Other postdoctoral researchers may have a very different experience. For example, the relatively small percentage of postdoctoral researchers working in national laboratories (including the NIH and other publically-funded research institutions) and in industry tend to earn more, have shorter appointment periods, and receive training and guidance with direct relevance to their career aspirations. Although, undoubtedly, there are many postdoctoral researchers at universities who gain valuable research experiences and receive useful mentoring to fulfill their career aspirations, this is not the case for a large number of postdocs, and the committee finds a need for significant reform. For this reason the recommendations that follow are intended to address the problems primarily encountered by postdoctoral researchers in the academic setting.

Using a definition of a postdoctoral researcher agreed upon by the NPA, NIH, and NSF as a guide—"An individual who has received a doctoral degree (or equivalent) and is engaged in a temporary and defined period of mentored advanced training to enhance the professional skills and research independence needed to pursue his or her chosen career path"—the committee has developed recommendations for best practices covering five aspects of the postdoctoral experience: period of service, title and role, career development, compensation and benefits, and mentoring. In addition, the committee stresses the importance of data collection through a sixth recommendation. While the recommendations are numbered, this is for ease of reference and should not be taken to imply prioritization; these six items are necessarily interconnected.

1. *Period of Service*: The committee endorses the recommended practice, put forward by the National Institutes of Health, the National Science Foundation, and the National Postdoctoral Association in 2007, that postdoctoral research training is and should be a "temporary and defined period." **Postdoctoral appointments for a given postdoctoral researcher should total no more than 5 years in duration, barring extraordinary circumstances.** This maximum term should include cumulative postdoctoral research experience, though extensions may be granted in extraordinary circumstances (e.g. family leave, illness).

This recommendation requires direct actions by the host institutions and the funding agencies.

1.1 Host institutions should maintain a record of how long a postdoctoral researcher remains in a position and provide that information to funding agencies as part of grant proposals.

1.2 To facilitate tracking of postdoctoral researchers, funding agencies could assign each postdoctoral researcher an identifier and keep a record of the total length of time any given individual is holding such a position.

2. *Title and Role*: In many instances, positions currently occupied by postdoctoral researchers are more appropriately filled by permanent staff scientists (e.g., technicians, research assistant professors, staff scientists, laboratory managers). **The title of "postdoctoral researcher" should be applied only to those people who are receiving advanced training in research.** When the appointment period is completed, the postdoctoral researchers should move on to a permanent position externally or be transitioned internally to a staff position with a different and appropriate designation and salary.

This recommendation requires action primarily by the funding agencies and the host institutions.

2.1 Funding agencies should have a consistent designation for "postdoctoral researchers," and require evidence that advanced research training is a component of the postdoctoral experience.

2.2 Host institutions should create or identity professional positions for individuals who are conducting research but who are not receiving training, and these individuals should receive appropriate remuneration, benefits, and privileges.

3. *Career Development*: **Host institutions and mentors should, beginning at the first year of graduate school, make graduate students aware of the wide variety of career paths available for Ph.D. recipients, and explain that postdoctoral positions are intended only for those seeking advanced research training.** Career guidance should include, where feasible, the provision of internships and other practical experiences. **The postdoctoral position should not be viewed by graduate students or principal investigators as the default step after the completion of doctoral training.**

This recommendation requires action by all the different members of the research system: the funding agencies, the host institutions, the professional societies, the mentors, the postdoctoral researchers, and even the graduate students before becoming postdoctoral researchers.

3.1 Host institutions, especially those with graduate student populations, should provide multiple engagement activities to help students explore all avenues of career development. Funding agencies should help to support these efforts.

3.3 Professional societies should gather and disseminate information about the full range of career paths within their discipline. Useful activities could include collecting statistics about job openings and salaries, identifying individuals in various sectors who can provide career advice, and organizing career fairs at professional meetings.

3.3 Mentors, in addition to providing guidance based on their own experience, should become familiar with and disseminate information about all forms of career development opportunities available either at the host institution or through their professional society.

3.4 Postdoctoral researchers and graduate students have a responsibility to participate in the career development opportunities provided by their institutions, to explore other sources of information such as professional societies, and to use available career-development tools.

4. ***Compensation and Benefits of Employment***: Current postdoctoral salaries are low. Salaries should be increased to (1) reflect the

qualifications of postdoctoral scholars, (2) address the slow progress the community has made toward implementing salary increases as recommended in several National Research Council reports, and (3) adjust the relative wage of postdoctoral researchers to appropriately reflect their value and contribution to research. The committee considered five different approaches for determining an appropriate minimum salary: (1) indexing to contemporary college graduates, (2) indexing to graduate stipends, (3) indexing to newly hired assistant professors, (4) inflation of previous recommendations, and (5) Research Grade Evaluation Guide. All of these approaches, which are discussed in detail in Appendix B, suggest an amount of $50,000 or more. In addition, despite considerable variation in salaries by field, geographic area, and sector, data on starting postdoctoral salaries reveal that the starting salary prescribed by the National Institutes of Health (NIH) for the Ruth L. Kirschstein National Research Service Award (NRSA) postdoctoral award (currently set at $42,000 for 2014) has become the *de facto* standard for many disciplines and on many academic campuses. **The NIH should raise the NRSA postdoctoral starting salary to $50,000 (2014 dollars), and adjust it annually for inflation. Postdoctoral salaries should be appropriately higher where regional cost of living, disciplinary norms, and institutional or sector salary scales dictate higher salaries.**[39]

In addition, host institutions should provide benefits to postdoctoral researchers that are appropriate to their level of experience and commensurate with benefits given to equivalent full-time employees. Comprehensive benefits should include health insurance, family and parental leave, and access to a retirement plan.

This recommendation requires action primarily by the funding agencies, with additional actions by the host institutions and the professional societies.

---

[39] Two of the committee members do not support the recommendation for a prescriptive "salary standard" based upon one particular field and funding agency (here, the National Institutes of Health [NIH] and life sciences) for two reasons: first, salaries—not just postdoctoral salaries—differ so much by discipline, region, funding agency, and type of institution (for example, the 2012 National Postdoctoral Association report indicates that about half of the institutions have minimum salaries that are lower than the 2013 NIH minimum of $39K; NPA 2012), and second, this "salary standard", meant to reflect a reasonable salary, will likely be used as a minimum salary. While they believe that institutions need flexibility to accommodate particular circumstances, they also firmly believe that a postdoctoral researcher's salary should be fair and fit rationally within the spectrum of salaries for researchers in that discipline, at that institution: for example, well above that of a graduate student and significantly less than that of an entry-level, career-track researcher, that is, permanent staff scientist, research track assistant professor, or tenure-track assistant professor.

4.1 Federal agencies should require host institutions to provide documentation of the salary a postdoctoral researcher will receive with all grant proposals.
4.2 Professional societies should collect data on salaries for all positions and make these publicly available.

5. *Mentoring*: Mentoring is an essential component of the postdoctoral experience and entails more than simply supervision. Mentoring should not be solely a responsibility of the principal investigator, although he or she should be actively engaged in mentoring. **Host institutions should create provisions that encourage postdoctoral researchers to seek advice, either formally or informally, from multiple advisors, in addition to their immediate supervisor. Host institutions and funding agencies should take responsibility for ensuring the quality of mentoring through evaluation of, and training programs for, the mentors.**

This recommendation requires action by the funding agencies and the host institutions, with supporting actions by the professional societies, the mentors, and the postdoctoral researchers themselves.
5.1 In addition to providing mentorship training and guidance to the immediate supervisors of the postdoctoral researchers, host institutions should establish mechanisms that make it easy for postdoctoral researchers to seek guidance from additional faculty or senior professionals who can enrich the postdoctoral training experience.
5.2 Funding agencies should identify better ways of evaluating or rewarding mentoring as an essential component of research. This could include mandatory self-reporting by mentors as well as blinded assessments by the postdoctoral researchers.
5.3 Professional societies are in an ideal position to provide additional mentors to supplement those at a postdoctoral researcher's host institution. This would be of particular value to postdoctoral researchers considering major career shifts such as a move from academia to industry.
5.4 Postdoctoral researchers need to recognize that a great research investigator is not necessarily equivalent to a great mentor and that many if not most principal investigators or senior research faculty have not received any formal training in mentoring. Therefore, postdoctoral researchers should seek guidance from a variety of people, and should be encouraged to do so.

6. ***Data Collection***: Current data on the postdoctoral population, in terms of demographics, career aspirations, and career outcomes are neither

adequate nor timely. Every institution that employs postdoctoral researchers should collect data on the number of currently employed postdoctoral researchers and where they go after completion of their research training, and should make this information publicly available. **The National Science Foundation should serve as the primary curator for establishing and updating a database system that tracks postdoctoral researchers, including non-academic and foreign-trained postdoctoral researchers.** Host institutions and federal agencies should cooperate with NSF on the data collection and maintenance process. Federal agencies and research institutions that report these data to the NSF should take advantage of various technologies that have become available in recent years to assist in timely and thorough collection.

Recognizing that this recommendation on data collection has been made many times before with little effect, the committee stresses that research institutions and professional societies should explore what they can do to enrich what is known about postdoctoral researchers and that all institutions make better use of new technologies and social and professional networks to collect relevant and timely data.

This recommendation requires action primarily by the funding agencies, with additional actions by the host institutions and the professional societies.

6.1 Funding agencies must improve their data collection on the postdoctoral segment of the workforce. This is especially true for the NSF, given its congressional mandate to "collect, acquire, analyze, report, and disseminate statistical data related to the science and engineering enterprise in the United States and other nations that is relevant and useful to practitioners, researchers, policymakers, and the public, including statistical data on research and development trends, [and] the science and engineering workforce… " (Section 505 of the America COMPETES Reauthorization Act of 2010). The NSF should work with other research agencies, particularly the NIH, to develop more reliable means of collecting data on postdoctoral researchers during and after their appointments. The use of a common identifier system for each postdoctoral researcher is a possible approach.

6.2 Host institutions should assist in the data collection efforts by remaining consistent with their labeling of postdoctoral researcher, keeping track of new hires and departures, and conducting exit interviews to determine career outcomes of their postdoctoral population. This information should be made publically available, particularly to prospective postdoctoral researchers.

6.3 Funding agencies should look favorably on grant proposals that include outcome data for an institution's postdoctoral researchers.

6.4 Professional societies should utilize their networks to collect information about career paths of their members and make this data easily available.

All of the reforms recommended here should be coordinated through a strong and separate or stand-alone postdoctoral office (PDO) at each host institution. These offices have become much more common since the publication of the 2000 Postdoctoral Report, and many have become members of the National Postdoctoral Association. However, more work is needed to truly enrich the postdoctoral experience. PDOs need to continue sharing experiences to help one another fulfill their potential to train mentors, organize career development activities, be a one-stop source of information for domestic and international postdoctoral researchers, manage postdoctoral researcher grievances, oversee data-gathering efforts, monitor institutional compliance with salary and benefits policy, and track the career progress of former postdoctoral researchers. Although currently these offices are often embedded within a larger graduate student affairs operation, they are essential for improving the visibility and recognition of postdoctoral researchers in their host institutions and deserve specialized recognition.

POTENTIAL IMPLEMENTATION AND BEST PRACTICES

The recommendations put forward by the committee define five aspects of the postdoctoral experience. Although postdoctoral researchers play a key role in the science and engineering enterprise, they are only one part of an increasingly complex system. All participants in this system can take directed and concrete steps towards the implementation of this vision for a better postdoctoral experience. This section outlines some potential outcomes, inspired by many of the best practices already implemented throughout the United States and around the world.

Given its complexity, it is important to approach the system holistically, as no single segment of the science and engineering enterprise can induce change on its own. Therefore, while the following potential outcomes and best practices are arranged by principal actor, many overlap in who would be involved.

Graduate Students

Ideally, doctoral students would give careful consideration to whether advanced research training in a postdoctoral position is required to further their career goals. They would seek information about the variety of career options early and often in their doctoral training. In addition to utilizing regular mentoring, graduate students would take full advantage of institutional and local resources that provide career development services.

Postdoctoral Researchers

Similarly to graduate students, postdoctoral researchers would ideally make repeated, realistic, and critical self-evaluations before, during, and after their postdoctoral experience concerning their career choices. They would take advantage of every opportunity for career planning, including, for example, the creation of an individual development plan. Postdoctoral researchers would not limit their focus solely to academic careers. To that end, they would seek advice and information from a variety of different sources, including their mentors and institutions, professional societies, and peers.

Mentors

Mentors and postdoctoral supervisors serve a particularly critical role in the science and engineering enterprise. With respect to postdoctoral researchers, mentors would recognize that the postdoctoral period should be viewed as a training period, and consequently that their role is to help individuals develop the necessary writing, laboratory management and leadership, communications, and other essential career-related skills. In most instances this will be best accomplished by a formal training program. However, it must be recognized that not all skills can be learned within the laboratory environment, especially those relating to non-research careers. Therefore, mentors, with the assistance of their institutions, would also provide postdoctoral researchers with substantial protected time to pursue career development activities.

In addition, because of the ever-increasing globalization of the science and engineering enterprise, mentors would be attuned to the special needs of temporary visa holders pursuing postdoctoral research, and consult with or provide referrals to experts within their institutions, including international offices.

Every postdoctoral researcher would have an individual development plan that is created with a mentor and reviewed yearly by someone in addition to the postdoctoral researcher's mentor (i.e., the head of the school or department or research division, or by the postdoctoral researcher's advisory committee, or by a specially appointed director of postdoctoral affairs). Similarly, institutions would encourage the establishment of advisory mechanisms to enable postdoctoral researchers to gain mentoring from a number of sources to complement the work of the primary mentor.

Institutions

Every institution would have at least one office or unit designated as responsible for the postdoctoral experience, policies, and activities, beyond that provided by the mentors. Ideally, there would be an independent office of postdoctoral affairs. Every effort would be made to provide postdoctoral researchers with the same type of recognition given to undergraduate and graduate students, faculty, and staff. The designated office would be responsible for collecting and maintaining statistics on the postdoctoral community within the institution, including long-term career outcomes. Institutions would make this information publicly available.

Like graduate and undergraduate students, postdoctoral researchers would receive an orientation upon arrival at their institution. This would include topics relating to safety, ethics, human resources, and other essential training as needed for the research discipline. In addition, postdoctoral researchers would receive an appointment letter that provides clear information and expectations about salary, benefits, duration of service, process for termination or resignation, protected time for career development, and intellectual property rights. Institutions would create formal and neutral grievance procedures to address conflicts between postdoctoral researchers and their direct supervisors. This procedure would also be identified in the appointment letter.

Institutions would invest resources to provide postdoctoral researchers and graduate students with information concerning the wide range of career opportunities. Where feasible, opportunities for practical experiences in other settings, such as teaching and both research- and non-research-based nonacademic employment, would be made available. Wherever possible, these career development activities would include internships for postdoctoral researchers and graduate students.

Above all, institutions would track, provide services, and have similar policies and procedures for postdoctoral researchers regardless of their source of funding.

Funders

All funding agencies would report annually to the National Science Foundation the number of postdoctoral researchers they have supported by discipline, visa status, degree-granting institution, and types of support. The NSF, through its National Center for Science and Engineering Statistics, would thereby track the number of postdoctoral researchers (according to the current agreed-upon definition) and follow their career outcomes in the same way as is currently done for Ph.D. recipients.

In addition, and because of the critical role of mentoring in the science and engineering enterprise, all funding agencies would place an emphasis on mentoring as a key criterion in evaluating grant proposals and the performance of principal investigators.

Professional Societies

Professional societies would recognize postdoctoral researchers as a distinct class of membership within their organizations and help postdoctoral researchers create a sense of community by facilitating postdoctoral researcher activities and networking at their meetings. They would involve postdoctoral researchers in the activities of their societies by promoting postdoctoral researcher service on committees, inviting postdoctoral researchers as speakers, and having postdoctoral researchers help to organize meetings.

Professional societies would provide postdoctoral researchers with career information and counseling similar to what they provide for graduate students. To this end, professional societies would help make broadly available information about job markets, career trajectories, and salaries for postdoctoral researchers and graduate students in their disciplines (e.g., through bulletins, or special sessions about career opportunities at meetings). Where possible, professional societies would collect, analyze, and publicize related information such as statistics about the numbers and kinds of job postings.

References

Alberts, Bruce, Marc W. Kirschner, Shirley Tilghman, and Harold Varmus. "Rescuing US biomedical research from its systemic flaws," *PNAS* (Perspective) 111(16): 5773-5777 April 22, 2014., doi:10.1073/pnas.1404402111. (Alberts 2014)

America COMPETES Act, Public Law 110-69 (Aug. 9, 2007; 110th Congress, 2007–2009), http://www.gpo.gov/fdsys/pkg/PLAW-110publ69/html/PLAW-110publ69.htm. (ACA 2007)

American Association of Universities. "Committee on Postdoctoral Education: Report and Recommendations." Washington, DC, 1998.

———. "Graduate and Postdoctoral Education Committee: Postdoctoral Education Survey Summary of Results." Washington, DC, 2005. (AAU 2005)

American Chemical Society. "Advancing Graduate Education in the Chemical Sciences: Summary Report of an ACS Presidential Commission." Washington, DC, 2013. (ACS 2013)

Assmus, Alexi. "The Creation of Postdoctoral Fellowships and the Siting of American Scientific Research." *Minerva* 31, no. 2 (1993): 151–183. (Assmus 1993)

Association of American Medical Colleges. "Compact Between Postdoctoral Appointees and Their Mentors." Washington, DC, 2006. (AAMC 2006)

Bhattacharjee, Yudhijit. "Postdoc Survey Finds Gender Split on Family Issues." *Science* vol. 318, Nov. 9, 2007.

Black, Grant, and Paula Stephan. "The Economics of University Science and the Role of Foreign Graduate Students and Postdoctoral Scholars." In *American Universities in a Global Market*, 129–161. Cambridge, MA: NBER Books, 2008.

Bonetta, Laura. "The Postdoc Experience: Taking a Long Term View." Postdoc Survey, doi: 10.1126/science.opms.r1000093. www.sciencecareers.org, AAAS/Science Business Office Feature, 2010.

Brown, Ryan. "Postdoctoral Fellowships in the Humanities Gain Importance in Career Paths." *The Chronicle of Higher Education*, August 28, 2011.

Burroughs Wellcome Fund and Howard Hughes Medical Institute. "Making the Right Moves: A Practical Guide to Scientific Management for Postdocs and New Faculty, Second Edition." Research Triangle Park, NC, and Chevy Chase, MD, 2006. (BWF 2006)

Cantwell, Brendan, and Jenny Lee. "Unseen Workers in the Academic Factory: Perceptions of Neoracism among International Postdocs in the United States and the United Kingdom." *Harvard Educational Review*, Winter 2010. (Cantwell and Lee 2010)

Cantwell, Brendan, and Barrett J. Taylor. "Internationalization of the postdoctorate in the United States: analyzing the demand for international postdoc labor." *High Education* 66:551–567, doi: 10.1007/s10734-013-9621-0, 2013.

Centre, Careers Research and Advisory. "Vitae Researcher Development Framework." Cambridge, UK, 2011.

Cheung, William W. L. "The Economics of Post-doc Publishing." *Ethics in Science and Environmental Politics*, doi: 10.3354/esep00083, vol. 8: 41–44, 2008.

Commission on Professionals in Science and Technology. "Postdocs: What We Know and What We Would Like to Know." Washington, DC, 2002. (CPST 2002)

Computing Research Association. "The Role of PostDocs in Computer Science : A Computing Research Association Working Paper." http://cra.org/postdocs/Issues-PostDoc-1-28-2011.pdf, 2011. Accessed April 10, 2014. (CRA 2011)

Corley, E., and M. Sabharwal. "Foreign-Born Academic Scientists and Engineers: Producing More and Getting Less Than Their U.S.-Born Peers?" *Research in Higher Education* 48, no. 8 (2007): 909–40. (Corley and Sabharwal 2007)

The J. David Gladstone Institutes. "Postdoctoral Fellow & Research Science Survey – Summary Report," 2008.

Davis, Geoff. "Doctors Without Orders." *American Scientist* 93, no. 3 supplement (2005). (Davis 2005)

———. "Improving the Postdoctoral Experience: An Empirical Approach.". In *The Science and Engineering Workforce in the United States*, Richard Freeman and Daniel Goroff, eds. Chicago, IL: NBER/University of Chicago Press, 2006.

Ehm, Kathleen Flint, and Cathee Johnson Phillips. "From Ph.D. To Professoriate: The Role of the Institution in Fostering the Advancement of Postdoc Women." Washington, DC: National Postdoctoral Association, 2013.

Einaudi, Peter, Ruth Heuer, and Patricia Green. "Counts of Postdoctoral Appointees in Science, Engineering, and Health Rise with Reporting Improvements" National Science Foundation, National Center for Science and Engineering Statistics. (NSF 13-334)Arlington, VA. September 2013 (Einaudi 2013)

Gallagher, Michael. Unpublished data presented at the "Alexander Humboldt Foundation's 6th Forum on the Internationalization of Sciences and Humanities, 21–23 October 2012, Berlin." (Gallagher 2012)

Garrison, Howard, Andrea Stith, and Susan Gerbi. "Foreign Postdocs: The Changing Face of Biomedical Science in the U.S." *The FASEB Journal* 19, no. 14 (2005): 1938–42.

Gibbs, Jr., Kenneth D., and Kimberly A. Griffin. "What Do I Want to Be with My Ph.D.? The Roles of Personal Values and Structural Dynamics in Shaping the Career Interests of Recent Biomedical Science Ph.D. Graduates." *CBE—Life Sciences Education* vol. 12, 711–723,Winter 2013.

Herzberg, F. "One more time: How do you motivate employees?" *Harvard Business Review*, 46, January-February, 53-62 (2003). (Herzberg 2003)

Jayasingam, Sharmila, and Jing Ren Yong. "Affective commitment among knowledge workers: the role of pay satisfaction and organization career management." *The International Journal of Human Resource Management* (2013) 24, 20, 3903 (Jayasingam 2013)

Kahn, Shulamit and Donna K. Ginther "Postdocs and Career Outcomes of Biomedical Ph.D.'s." Presentation to COSEPUP Ad-Hoc Committee on Postdocs, December 13, 2011. (Kahn 2011)

Leyman, Annik. "A Postdoc: What Is It Worth Outside Academia?" Vitae Researcher Development International Conference, Abstracts, 2011.

Macalister, Heather E. "Postdoctoral Training Reform: A Program for Scientist-Teachers." Science Careers. http://sciencecareers.sciencemag.org.

Martinez, Elisabeth D., et al. "Falling off the academic bandwagon." EMBO reports vol. 8, no. 11, 2007.

Misu, Toshiyuki, Tomoko Shimomura, Yukiko Miura, Akira Horoiwa, and Kan Imai. "Survey on Postdoctoral Fellows and Research Assistants (FY2005)." 1st Policy-Oriented Research Group National Institute of Science and Technology Policy (NISTEP) in cooperation with Knowledge Infrastructure Policy Division Science and Technology Policy Bureau Ministry of Education, Culture, Sports, Science and Technology (MEXT). Japan. August 2006. Available at http://www.nistep.go.jp/achiev/ftx/eng/mat128e/idx128e.html. Accessed on May 8, 2014.

Misu, Toshiyuki and Akira Horoiwa "Japanese postdocs seek their path." *Nature* 451(742) doi: 10.1038/nj7179-742b (2008).

MIT Institutional Research. "Postdoctoral Life at MIT: Findings from the 2010 Postdoctoral Scholar Survey," 2011. http://web.mit.edu/IR/surveys/pdf/Postdoctoral_Life_at_MIT_Report_June_2011.pdf. Accessed April 10, 2014.

Mitchell, Jeremy, and Valerie Walker. "The 2013 Canadian Postdoc Survey: Painting a Picture of Canadian Postdoctoral Scholars." The Canadian Association of Postdoctoral Scholars/L'Association Canadienne de Stagiaires Post-doctoraux (CAPS-ACSP) and Mitacs, 2013.

MOR Associates, Inc. "Berkeley Lab 2009 Postdoc Survey." Brian McDonald, President, March 2009.

National Academy of Sciences, National Academy of Engineering, Institute of Medicine. "Adviser, Teacher, Role Model, Friend: On Being a Mentor to Students in Science and Engineering." Washington, DC: The National Academies Press, 1997.

———. "Enhancing the Postdoctoral Experience for Scientists and Engineers: A Guide for Postdoctoral Scholars, Advisers, Institutions, Funding Organizations, and Disciplinary Societies." Washington, DC: The National Academies Press, 2000. (The 2000 Postdoctoral Report)

National Institutes of Health. "NIH Statement in Response to the NAS Report: Addressing the Nation's Changing Needs for Biomedical and Behavioral Scientists." March 22, 2001. NOT-OD-01-027 http://grants1.nih.gov/grants/guide/notice-files/NOT-OD-01-027.html (NIH 2001)

———. "NIH Intramural Research Postdoctoral Fellows Mentoring Survey Report." NIH Fellows Committee Mentoring Subcommittee, Office of Intramural Research, 2011. (NIH 2011)

———. "Biomedical Research Workforce Working Group Report." Bethesda, MD, 2012. (BMW 2012)

———. "NIH Postdoc Handbook." Bethesda, MD: Office of Intramural Training and Education, 2012.

National Postdoctoral Association. "National Postdoctoral Association Institutional Survey on Postdoctoral Compensation, Benefits, and Professional Development Opportunities." Released April 2012. http://www.nationalpostdoc.org/images/stories/ Documents/Other/npa-survey-report-april-2012.pdf. Accessed April 9, 2014. (NPA 2012)

National Research Council. "Invisible University: Postdoctoral Education in the United States. Report of a Study Conducted Under the Auspices of the National Research Council. [Richard B. Curtis, Study Director]." Washington, DC: The National Academies Press, 1969. (NRC 1969)

———. "Postdoctoral Appointments and Disappointments." Washington, DC: National Academy Press, 1981. (NRC 1981)

———. "Trends in the Early Careers of Life Scientists." Washington, DC: The National Academies Press, 1998. (NRC 1998)

———. "Addressing the Nation's Changing Needs for Biomedical and Behavioral Scientists." Washington, DC: The National Academies Press, 2000. (NRC 2000)

———. "Gender Differences at Critical Transitions in the Careers of Science, Engineering, and Mathematics Faculty." Washington, DC: The National Academies Press, 2000. (NRC 2000)

———. "Bridges to Independence: Fostering the Independence of New Investigators in Biomedical Research." Washington, DC: The National Academies Press, 2005. (NRC 2005a)

———. "Policy Implications of International Graduate Students and Postdoctoral Scholars in the United States." Washington, DC: The

National Academies Press, 2005. (NRC 2005b)

———. "Gender Differences at Critical Transitions in the Careers of Science, Engineering, and Mathematics Faculty." Washington, DC: The National Academies Press, 2010. (NRC 2010)

———. "Research Training in the Biomedical, Behavioral, and Clinical Research Sciences." Washington, DC: The National Academies Press, 2011. (NRC 2011)

National Science Board Subcommittee on Science and Engineering Indicators. "Science and Engineering Indicators 2008." Arlington, VA (NSB 08-01; NSB 08-01A), January 2008. (NSB 2008)

National Science Board Subcommittee on Science and Engineering Indicators. "Science and Engineering Indicators 2014" Arlington, VA (NSB 14-01), February 2014.

The National Science Foundation. "Proposal and Award Policies and Procedures Guide: Part I - Proposal Preparation & Submission Guidelines (GPG)." Effective June 1, 2007. NSF 07-140. OMB Control Number: 3145-0058. April 2007.

National Science Foundation, Division of Science Resources Statistics. "Postdoc Participation of Science, Engineering, and Health Doctorate Recipients." Arlington, VA (NSF 08-307), 2008. (NSF 2008)

Oklahoma State University, Office of Institutional Research and Information Management. "2013-2014 Faculty Salary Survey of Institutions Belonging to Association of Public and Land-Grant Universities." Stillwater, OK (2014). (OSU 2014).

Pennsylvania State University. "Postdoctoral General Status Survey Summary – October 2008," 2008.

President's Council of Advisors on Science and Technology. "Engage to Excel: Producing One Million Additional College Graduates with Degrees in Science, Technology, Engineering, and Mathematics." Report to the President Executive, U.S. Office of Science and Technology Policy, Office of the President, 2012. (PCAST 2012a)

———. "Transformation and Opportunity: The Future of the U.S. Research Enterprise." Report to the President Executive, U.S. Office of Science and Technology Policy, Office of the President, 2012. (PCAST 2012b)

Puljak, Livia, and Wallace D Sharif. "Postdocs' perceptions of work environment and career prospects at a US academic institution." *Research Evaluation*, 18(5): 411–415, December 2009, doi:10.3152/095820209X483064.

Roberts, Gareth. "SET for success: The supply of people with science, technology, engineering and mathematics skills." The report of Sir Gareth Roberts' Review, April 2002. http://webarchive.nationalarchives.gov.uk/+/http:/www.hm-treasury.gov.uk/d/robertsreview_introch1.pdf. Accessed April 16, 2014. (Roberts 2002)

Romer, Paul M. "Should the Government Subsidize Supply or Demand in the Market for Scientists and Engineers?" Ch. 7 in *Innovation Policy and the Economy*, vol. 1 (out-of-print volume from the National Bureau of Economic Research; http://www.nber.org/books/jaff01-1, accessed April 9, 2014), Adam B. Jaffe, Josh Lerner, and Scott Stern, eds. MIT Press, 2001. (Romer 2001)

Sauermann, Henry, and Michael Roach. "Science Ph.D. Career Preferences: Levels, Changes, and Advisor Encouragement." PLoS ONE 7(5):e36307, doi:10.1371/journal.pone.0036307, 2012. (Sauermann and Roach 2012)

———. "Junior S&E's and Academic Research Careers – Insights from the Science and Engineering Ph.D. and Postdoc Survey (SEPPS) 2010/2013." Presentation to COSEPUP. (Sauermann and Roach 2013)

SigmaXi and National Postdoctoral Association. "Professionalizing the Postdoctoral Experience." Research Triangle Park, NC, 2006. (SigmaXi 2006)

Stephan, Paula. "How Economics Shapes Science." Cambridge, MA: Harvard University Press, 2012. (Stephan 2012)

———. "The Endless Frontier: Reaping What Bush Sowed?" NBER Working Paper 19687. http://www.nber.org/papers/w19687. Accessed April 10, 2014. (Stephan 2013)

Stephan, Paula, Chiara Franzoni, and Giuseppe Scellato. "Choice of Country by the Foreign Born for Ph.D. and Postdoctoral Study: A Sixteen-Country Perspective." (2013): NBER Working Paper 18809, February. (Stephan, Franzoni, and Scellato 2012)

Stephan, Paula, and Jennifer Ma. "The Increased Frequency and Duration of the Postdoctorate Career Stage." The *American Economic Review*, vol. 95, no. 2, Papers and Proceedings of the One Hundred Seventeenth Annual Meeting of the American Economic Association, Philadelphia, PA, January 7–9, 2005 (2005): 71–75. (Stephan and Ma 2005)

Sukthankar, Ashwini. "Visas, Inc: Corporate Control and Policy Incoherence in the U.S. Temporary Foreign Labor System." Brooklyn, NY: Global Workers Justice Alliance, 2012.

Tabak, Lawrence A. and Francis S. Collins. "Weaving a Richer Tapestry in Biomedical Science" *Science*. 333(6045): 940–941 Aug 19, 2011. (Tabak and Collins 2011)

Thatcher, J.B., Liu, Y., Stepina, L.P. Goodman, J.M. Treadway, D.C. "IT worker Turnover: An empirical investigation of Intrinsic Motivation" The DATA BASE for Advances in Information Systems - Spring-Summer pp. 133 – 146 (2006). (Thatcher et at. 2006)

Thompkins, Sibby Anderson, Vanessa González-Pérez, and Jennifer Cohen. "Changing the Culture of Science: Minority Postdoctoral Organizations Help lead the Way." *The POSTDOCket* vol. 11, iss. 3, Summer 2013.

Unknown Author. "History of Extramural Research and Training Programs at NIH," mimeo, circa 1959.

U.S. Office of Personnel Management. "Research Grade Evaluation Guide." September 2006. Available at http://www.opm.gov/policy-data-oversight/classification-qualifications/classifying-general-schedule-positions/functional-guides/gsresch.pdf (RGEG 2006)

Virginia Commonwealth University. "Fall 2009 Postdoctoral Survey Summary," 2009.

Vitae. "Concordat to Support the Career Development of Researchers: An Agreement between the Funders and Employers of Researchers in the UK" https://www.vitae.ac.uk/policy/vitae-concordat-vitae-2011.pdf. Accessed May 8, 2014

von Bubnoff, Andreas. "Superpostdocs Reach for the Stars." *Cell*, vol. 128, iss. 6 (March 23, 2007): 1023–1026. http://dx.doi.org/10.1016/j.cell.2007.03.005. (von Bubnoff 2007)

Weinberg Bruce A., Jason Owen-Smith, Rebecca F. Rosen, Lou Schwarz, Barbara McFadden Allen, Roy E. Weiss, and Julia Lane. "Science Funding and Short-Term Economic Activity." *Science* Vol. 344 no. 6179 (April 2014). (Weinberg 2014)

Wolfe, Benjamin, Natacha Bodenhausen, and Luciana Molinero. "Results and Analysis of the 2008 Biological Sciences Division Postdoctoral Association Survey." Presentation for the University of Chicago.

Wasem, Ruth Ellen. "Immigration of Foreign Nationals with Science, Technology, Engineering, and Mathematics (STEM) Degrees." Congressional Research Service, 7-5700. www.crs.gov. R42530, 2012. (CRS 2012)

Zielinska, Edyta, Dan Cossins, and Beth Marie Mole. "The Scientist Best Places to Work Postdocs, 2013" Scientist, 27 (4): 48-51, April 2013. (*Scientist* 2013)

Appendixes

APPENDIX A:
UNIQUE CHALLENGES OF INTERNATIONAL POSTDOCTORAL RESEARCHERS IN THE UNITED STATES

The preponderance of international postdoctoral researchers working in the United States underscores the crucial role that these talented researchers play in the overall research endeavor. In addition to many of the usual challenges confronted by postdoctoral researchers in general, international postdoctoral researchers face unique circumstances that can also create problems related to language, culture, law, and visas.

Visas

Foremost among these challenges are visa issues and the related restrictions on travel outside the United States. Visa requirements also present challenges for international postdoctoral researchers in terms of restrictions on the employment of spouses and their own ability to remain in the country when moving from one position to another. In fact, these visa restrictions can place international postdoctoral researchers in a more vulnerable position with their principal investigators, because losing a position may make it impossible for the individual and spouse or family to remain in the country. Even the prospect of losing one's visa status may reduce the postdoc's ability or willingness to advocate for what would be normal workplace issues for domestic postdoctoral researchers.

One of the leading challenges facing international postdoctoral researchers is navigating the morass of visa regulations, requirements, and restrictions. The already complex regulations have been made even more difficult with the tightening of visa regulations in the wake of terrorist attacks in 2001. The current types of visas available to incoming international postdoctoral researchers are listed in Box A-1.

The EB (extraordinary ability) visas are very rare, and most postdoctoral researchers are on J-1 (exchange visitors) or H-1B (specialty occupations) visas. In the past decade, many universities and other institutions have created guidelines to help postdoctoral researchers, mentors, and administrators navigate through the complexities of visa procurement and maintenance. The National Postdoctoral Association (NPA) provides open access to "A Quick Guide to Visas" on its website and provides more in-depth information for its members. The public site provides an overview of visa types suitable for most postdoctoral researchers, a glossary of terms, and a list of frequently asked questions.

Box A-1
Types of Visas, Rules and Restrictions

- Non-immigrant, employer-sponsored
  - F-1 OPT: STEM Optional Practical Training Extension
    - 2-year maximum
    - F-2 spousal visa
  - J-1: Exchange Visitors
    - 5-year maximum followed by 2 years out of country
    - J-2 employable spousal visa
  - H-1B: Specialty Occupations
    - 6-year maximum
    - H-4 unemployable spousal visa
    - University employees not counted toward cap on total number of visas
- Immigrant, self-sponsored
  - E-1A: Extraordinary Ability
  - E-2: National Interest Waiver
- Immigrant, employer-sponsored
  - E-1B: Outstanding Researcher
  - E-2: Advanced Degree

SOURCES: More information about U.S. Visa rules and restrictions is available at http://travel.state.gov/content/visas/english.html. Accessed May 8, 2014.

The visa system is complex and often a mystery to mentors as well as postdoctoral researchers. Institutions that host foreign postdoctoral researchers would provide a valuable service by creating a central source of information guidance. Most research institutions have administrative offices dedicated to international students that can assist both postdoctoral researchers and their mentors navigate the complexities of applying for and maintaining appropriate visas. However, in many cases, these offices are not well versed in the unique needs of postdoctoral trainees, because their primary focus is undergraduate or graduate students. In addition, frequently there is little coordination between these offices and departments and individual mentors, either by direct assistance or by general education about visa requirements. Anecdotal evidence suggests that, all too often, important visa decisions are left in the hands of individual mentors who do not have sufficient understanding of the implications. Even well-meaning mentors or departmental administrators can place international postdoctoral researchers in future jeopardy if they do not have a complete understanding of the visa system. What is needed is close and continued cooperation between institutional offices, mentors, departmental administrators, and international postdoctoral researchers to establish guidelines and procedures to ensure visas are handled in the most appropriate manner.

English as Second Language and Culture Skills

Additional challenges faced by many international postdoctoral researchers result from language and cultural differences. Although most postdoctoral researchers for whom English is not a native language have acquired sufficient English language skills prior to arriving in the United States, or engage in formal and informal activities to improve communications skills, many still might need additional assistance to develop the language ability necessary to be an effective teacher or to work in industry.

Even with adequate language proficiency, international postdoctoral researchers face cultural challenges that may result in impediments to proper advising and career advancement. Questions of how to approach mentors and colleagues arise even for those raised in the same culture, but these are compounded for postdoctoral researchers from countries that have vastly different cultural norms for dealing with authority figures. In some cases, these cultural differences can lead to misunderstandings that impede adequate mentoring. On the other end of the spectrum, those international postdoctoral researchers who work in laboratories made up exclusively of colleagues from the same culture may not develop the requisite communication skills to be viable job candidates in the future. Postdoctoral researchers at some institutions have formed self-help groups to deal with some of these issues. That is certainly helpful, and, under best practices, institutions would facilitate and support such efforts.

International postdoctoral researchers also need help in dealing with U.S. institutions, such as public schools, the health care system, the motor vehicles department, and the Internal Revenue Service, all of which can be problematic even for U.S. natives.

The United States could also learn from what other countries are doing in postdoctoral training. U.S. institutions should be aware of salaries and benefits in other countries, as well as the quality of training and career prospects so that they can maintain a postdoctoral training system that can attract the most talented researchers. The United States can learn from innovative and effective policies developed in other countries. For example, the United Kingdom has an important document, "Concordat to Support the Career Development of Researchers," signed by funding agencies and prominent scientific institutions, and supported by universities and professional societies, that sets important standards for the role of individual scientists in the pursuit of research. Under the principles of the Concordat, both funders and researchers are obligated to regularly review their progress as to whether scientists under their purview are given the support called for by the agreement.

The creation of the NPA was an important initiative by postdoctoral researchers themselves to make information available and to advocate for postdoctoral interests. As the research enterprise becomes more global, it would be useful to organize postdoctoral researchers and the various postdoctoral associations across the globe to share information and promote the interests of postdoctoral researchers.

APPENDIX B:
DETERMINATION OF THE MINIMUM SALARY FIGURE

The low salaries of many postdoctoral researchers have long been a source of concern in terms of both fairness and the incentives that low salaries create for principal investigators to employ postdoctoral researchers. With the exception of positions in engineering and some parts of the physical sciences, beginning salaries for postdoctoral researchers in academia correlate closely with the National Institutes of Health's (NIH) minimum stipend for the Ruth L. Kirschstein National Research Service Award (NRSA). Even in the field of engineering, there is minimal difference between postdoctoral salaries and the NIH NRSA minimum. Salaries for postdoctoral researchers outside of academia are usually higher. The committee devoted considerable attention to determining an appropriate minimum and explored several approaches to finding corresponding salaries for comparison.

Indexing to contemporary college graduates

One benchmark is the salary earned by a college graduate who did not go to graduate or professional school and entered the marketplace six to seven years earlier. According to the U.S. Census Bureau, this salary was $49,911 in 2012.

Indexing to graduate student stipends

Data on graduate stipends are difficult to compile. However, using the crowdsourcing website GradPay and IPEDS Institutional Characteristics Survey Tuition Data, it was determined that the average total cost for a science, engineering, or health graduate student in 2011 was approximately $51,000, more than half of which is for tuition and fees.[40]

Indexing to newly hired assistant professors

According to the *2013-2014 Faculty Salary Survey of Institutions Belonging to Association of Public and Land-Grant Universities* conducted by Oklahoma State University, the average nine-month salary of a newly hired assistant professor at public research universities in the biological and biomedical sciences is $74,177; in engineering it is $84,012; in computer and information

[40] Values rounded to the nearest thousand. Data were gathered for 73 institutions and 51 self-identified science, engineering, or health fields. More information about GradPay is available at http://gradpay.herokuapp.com/, last accessed April 8, 2014.

sciences and support services it is $94,234 (OSU 2014). Assuming a reasonable starting salary for a postdoctoral researcher to be approximately two-thirds of this nine-month amount implies a starting salary in the biological and biomedical sciences for postdoctoral researchers of $49,698; in engineering, of $56,288; in computer and information sciences and support services of $63,137. The 67 percent average for all non-medical related disciplines is $52,062.

Inflation of previous recommendations

Previous reports—from the National Research Council and others—have put forward recommendations for minimum salaries. Recommendation 5.4 in the National Research Council' s *Addressing the Nation's Changing Needs for Biomedical and Behavioral Scientists* states "[s]tipends and other forms of compensation for those in training should be based on education and experience and should be regularly adjusted to reflect changes in the cost of living." (NRC 2000). In 2001, the NIH drafted a response to this recommendation, saying:

> The NIH concurs with the committee's observation that NRSA stipends are unduly low in view of the high level of education and professional skills involved in biomedical research. NRSA stipend targets have been identified for both predoctoral and postdoctoral recipients. NIH plans to develop budget requests that will permit an increase in stipends by 10 to 12 percent per year for the next few years in order to reach those targets. Once the targets have been attained, annual cost-of-living adjustments will be instituted so that stipend levels remain relatively constant in real value. The appropriateness of stipend levels will be examined in future years to maintain comparability to levels of income available to students and postdoctorates from other sources. (NIH 2001)

At the time of the NRC recommendation, entry level NRSA postdoctoral salaries were $26,256.[41] The NIH set, at that time, a "tentative target" of $45,000 for entry-level postdoctoral salaries. Although the NIH did not set a target date, using the proposed technique to increase the levels gradually, this goal could have been reached around 2005. Adjusted for inflation, a beginning postdoctoral salary of $45,000 in 2005 would be equivalent to approximately $54,846 in 2014.

Research Grade Evaluation Guide

The NIH and other federal agencies use the Research Grade Evaluation Guide (RGEG) to determine the correct placement of their intramural researchers on the General Schedule (GS) for salary (RGEG 2006). A typical

---

[41] More information about NRSA Stipend level can be found at http://grants.nih.gov/training/nrsa.htm, last accessed October 29, 2014.

beginning researcher would be placed at GS-11, which has a minimum starting salary of $50,790 in 2014. The GS system also makes adjustments for local cost of living. For example, the minimum for a GS-11 is $63,386 in Boston, MA, and $59,302 in Mesa, AZ.[42]

The $50,000 figure

Because all of these comparisons pointed to a starting salary of approximately $50,000 or more, the committee concluded that a conservative minimum should be $50,000. The committee recognizes that it cannot—and should not—dictate salaries to all institutions. However, the justifications for recommending this minimum for what has become the *de facto* benchmark for many institutions and disciplines—the NIH's NRSA starting salary level—are strong.

The committee emphasized that this is a benchmark minimum. A number of factors—including but not limited to discipline, region, and institutional salary scales—should be taken into account when setting salaries for postdoctoral researchers.

[42] The 2014 General Schedule (GS) Locality Pay Tables are available at http://www.opm.gov/policy-data-oversight/pay-leave/salaries-wages/2014/general-schedule/, last accessed on October 30, 2014.

APPENDIX C:
STAKEHOLDER SPEAKERS

The committee organized three public stakeholder sessions, including a panel at the 2012 National Postdoctoral Association's Annual Meeting. Testimonies were provided by the following individuals:

SATYAVEDA BHARATH, Postdoctoral Fellow, Department of Chemistry and Biochemistry, University of Maryland
LYNDA CARLSON, Director, National Center for Science and Engineering Statistics, National Science Foundation
LORI CONLAN, Director, Office of Postdoctoral Services, Office of Intramural Training & Education, National Institutes of Health
GEOFF DAVIS, Staff Researcher, Google; formerly Visiting Scholar, Sigma Xi
CHRISTINE DES JARLAIS, Assistant Dean for Postdoctoral Affairs and Diversity, University of California, San Francisco
DEBAPRIYA DUTTA, Counselor, Science and Technology, Embassy of India
ENO EBONG, Postdoctoral Fellow, City College of New York and Albert Einstein College of Medicine
ZOE FONSECA-KELLY, Senior Research Fellow, Massachusetts Eye and Ear Infirmary
GARTH FOWLER, Assistant Chair in Neurobiology Department, Northwestern University
ROBERT FRIEDMAN, Deputy Director, J. Craig Venter Institute, California
STACY GELHAUS, Former Chair, National Postdoctoral Association
MARCEL HAAS, Postdoctoral Research Scholar, Space Telescope Science Institute
JUSTIN HALL, Senior Scientist, Pfizer Pharmaceuticals, Inc.
MICHAEL HAYES, UAW Local 5810 Unit Chair and Postdoctoral Scholar, University of California, Irvine
SHULAMIT KAHN, Associate Professor of Finance and Economics, School of Management, Boston University
EBEN KIRKSEY, Mellon Postdoctoral Fellow, City University of New York
JAMES LIGHTBOURNE, Senior Advisor, National Science Foundation
SHARON METZGER, Senior Fellowship and Postdoctoral Scholar Program Analyst, University of California, Irvine

AURÉLIE NÉVÉOL, Research Fellow, National Library of Medicine, National Center for Biotechnology Information
ANNE PODUSKA, Senior Program Associate, American Association for the Advancement of Science
ALBERTO ROCA, Diversity Consultant, Recruiter, and Editor at MinorityPostdoc.org
SALLY ROCKEY, Deputy Director for Extramural Research, National Institutes of Health
MICHAEL TEITELBAUM, Senior Advisor, Alfred P. Sloan Foundation
RODNEY ULANE, Training Officer and Director, Division of Scientific Programs, NIH Office of Extramural Research, National Institutes of Health
MAX VÖGLER, Director, Deutsche Forschungsgemeinschaft North America Office
MARY ANNE WITH, Postdoc Program Office Leader, Los Alamos National Lab

APPENDIX D:
ACRONYMS

TABLE C-1 Acronyms used in the report.

AAAS	The American Association for the Advancement of Science
AAMC	The Association of American Medical Colleges
AAS	The American Astronomical Society
AAU	The Association of American Universities
ACA	America COMPETES Act
ACS	The American Chemical Society
AGU	The American Geophysical Union
APS	The American Physical Society
BMW Report	The Biomedical Research Workforce Working Group Report
BMW Working Group	The NIH Biomedical Research Workforce Working Group
BWF	The Burroughs Wellcome Fund
CAPS-ACSP	The Canadian Association of Postdoctoral Scholars/L'Association Canadienne de Stagiaires Post-doctoraux
COSEPUP	The Committee on Science, Engineering, and Public Policy
CPST	The Commission on Professionals in Science and Technology
CRA	The Computing Research Association
CRS	Congressional Research Service
D.Eng.	Doctorate of Engineering
DGE	The Division of Graduate Education
EHR	Education and Human Resources
FASEB	The Federation of American Societies for Experimental Biology
FFRDC	Federally funded research and development center
FY	Fiscal Year

GREAT	Graduate Research, Education, and Training
GSICE	The Graduate Student Internship for Career Exploration
GSS	The Survey of Graduate Students and Postdoctorates in Science and Engineering
HHMI	The Howard Hughes Medical Institute
IDP	Individual Development Plan
IEEE	The Institute of Electrical and Electronics Engineers
IOM	The Institute of Medicine
IPEDS	The Integrated Postsecondary Education Data System
MD	Medical Doctorate
MEXT	The Ministry of Education, Culture, Sports, Science and Technology
NAE	The National Academy of Engineering
NAS	The National Academy of Sciences
NIGMS	The National Institute of General Medical Sciences
NIH	The National Institutes of Health
NISTEP	The National Institute of Science and Technology Policy
NPA	The National Postdoctoral Association
NRC	The National Research Council
NRSA	National Research Service Award
NSB	The National Science Board
NSF	The National Science Foundation
OITE	The NIH Office of Intramural Training and Education
OPA	Office of Postdoctoral Affairs
PAESMEM	The Presidential Awards for Excellence in Science, Mathematics and Engineering Mentoring Program
PCAST	The President's Council of Advisors on Science and Technology
PDA	Postdoctoral Association
PDO	Postdoctoral Office
Ph.D.	Doctorate of Philosophy
Sc.D.	Doctorate of Science
SDR	The Survey of Doctorate Recipients
SED	The Survey of Earned Doctorates
SEPPS	The Science and Engineering Ph.D. and Postdoc Survey
STEM	Science, Technology, Engineering, and Mathematics

UC	The University of California
UCSF	The University of California, San Francisco
UNC	The University of North Carolina
URM	Underrepresented Minorities
USDA	U.S. Department of Agriculture

APPENDIX E: COMMITTEE MEMBER BIOGRAPHIES

Gregory A. Petsko, Ph.D.. (Chair) is the Arthur J. Mahon Professor of Neurology and Neuroscience at Weill Cornell Medical College and Director of the Helen and Robert Appel Alzheimer's Disease Research Institute. He is also Professor of Biomedical Engineering at Cornell University and Gyula and Katica Tauber Professor of Biochemistry and Chemistry, Emeritus, at Brandeis University where he also served as Director of the Rosenstiel Basic Medical Sciences Research Center and Chair of the Department of Biochemistry. He received his D. Phil. from Oxford University in molecular biophysics. His research interests include protein structure and function and the development of methods to treat age-related neurodegenerative diseases, including ALS, Alzheimer's and Parkinson's Diseases. His awards include the Lynen Medal, Sidhu Award of the American Crystallographic Association, Pfizer Award in Enzyme Chemistry of the American Chemical Society, an Alexander von Humboldt Senior Scientist Award, the Max Planck Prize, and a Guggenheim Fellowship. He is past-president of the American Society for Biochemistry and Molecular Biology, and President of the International Union of Biochemistry and Molecular Biology. He is an elected member of the American Academy of Arts and Sciences, American Philosophical Society, National Academy of Sciences, and the Institute of Medicine.

Sibby Anderson-Thompkins, Ph.D., is director of postdoctoral affairs at the University of North Carolina (UNC) at Chapel Hill. She earned a B.A. and an M.A. in communication studies from UNC and an M.S. in educational research and Ph.D. in educational policy studies from Georgia State University. Dr. Anderson-Thompkins served as assistant dean of students and assistant dean in the Office of Student Academic Counseling at UNC. She was also an associate dean of student affairs at Hampshire College and a clinical faculty member in education and dean of advising at Agnes Scott College. She is a former diversity officer of the National Postdoctoral Association (NPA) and serves on the advisory group for the NPA ADVANCE Project *From Postdoc to Faculty: Transition Issues for Women Scientists*.

H. Russell Bernard, Ph.D., is professor emeritus of anthropology at the University of Florida. He is the founder and current editor of the journal *Field Methods*, and has served as editor for the *American Anthropologist* and *Human*

Organization. He has also served as the chairman of the board of directors for the Human Relations Area Files. His teaching interests focus on research design and the systematic methods available for collecting and analyzing field data. He has taught both within the United States and in Greece, Japan, Germany, and England. Bernard received his B.A. in anthropology and sociology from Queens College, New York, his M.A. in anthropological linguistics from the University of Illinois, and his Ph.D. in anthropology from the University of Illinois. Bernard has been a recipient of the Franz Boas Award from the American Anthropological Association as well as the University of Florida Graduate Advisor/Mentoring Award. He is a member of the National Academy of Sciences.

Carol Greider, Ph.D., is the Daniel Nathans Professor and the director of molecular biology and genetics at the Johns Hopkins Institute of Basic Biomedical Sciences at Johns Hopkins University. Dr. Greider is co-recipient of the 2009 Nobel Prize in Physiology or Medicine for the discovery of how chromosomes are protected by telomeres and the enzyme telomerase, an enzyme that maintains the length and integrity of chromosome ends and is critical for the health and survival of all living cells and organisms. Dr. Greider graduated from the University of California, Santa Barbara, with a bachelor's degree in biology and earned her Ph.D. in molecular biology from the University of California, Berkeley. Greider has also won the Gardiner Award, Rosenstiel Award, Passano Foundation Award, Richard Lounsbery Award, Wiley Prize, and the Albert Lasker Award for Basic Medical Research. Dr. Greider is an elected member of the American Academy of Arts and Sciences, National Academy of Sciences, and the Institute of Medicine.

James Plummer, Ph.D., is Frederick Emmons Terman Dean of the School of Engineering and John M. Fluke Professor of Electrical Engineering at Stanford University. Dr. Plummer received a B.S. from the University of California, Los Angeles, and an M.S. and Ph.D. in electrical engineering from Stanford University. He serves on the board of directors and on the technical advisory boards of several companies and was one of the founders of T-RAM. Dr. Plummer is the recipient of the Andrew S. Grove Award, Aldert Van der Ziel Award, J.J. Ebers Award, and the IEEE Third Millennium Medal. He also received the IEDM Paul Rappaport Award, McGraw-Hill/Jacob Millman Award, Aviation Week & Space Technology 2003 Laurels Award for Electronics, and the University Research Award from the Semiconductor Industry Association. Dr. Plummer is a fellow of the Institute of Electrical and Electronics Engineers, and an elected member of the American Academy of Arts and Sciences and the National Academy of Engineering.

E. Albert Reece, M.D., Ph.D., M.B.A., is vice president for medical affairs at the University of Maryland, the John Z. and Akiko K. Bowers Distinguished

Professor, and dean of the School of Medicine. He also is a professor in the departments of Obstetrics, Gynecology and Reproductive Sciences; Internal Medicine; and Biochemistry & Molecular Biology, where he oversees an active multimillion dollar NIH-funded research program, studying the biologic and molecular causes and consequences of diabetes-induced birth defects. Dr. Reece received a B.S. degree with honors (Magna Cum Laude) from Long Island University, and an M.D. degree from New York University School of Medicine; completed his residency in OB/GYN at Columbia University Medical Center, and a fellowship in maternal-fetal medicine at Yale University School of Medicine. In addition, he has a Ph.D. degree in biochemistry from the University of the West Indies, Kingston, Jamaica; and an M.B.A. degree from the Fox School of Business & Management at Temple University in Philadelphia, Pennsylvania. He was a member of the Yale faculty for almost 10 years, before being recruited to serve as the Abraham Roth Professor and Chair of Obstetrics and Gynecology at the Temple University School of Medicine. Prior to joining the University of Maryland School of Medicine, he served as vice chancellor and dean of the University of Arkansas College of Medicine. He is a member of the prestigious Institute of Medicine (IOM) of the National Academy of Sciences, and recently served as chair of the Association of American Medical College's Council of Deans.

Nancy Schwartz, Ph.D., is professor of pediatrics, biochemistry, and molecular biology at the University of Chicago. Dr. Schwartz is also dean for postdoctoral affairs, having served as founding dean of graduate and postdoctoral affairs in the Division of Biological Sciences for 25 years. She earned her B.S. in chemistry and her Ph.D. in biochemistry from the University of Pittsburgh. Her research centers on the role of extracellular matrix components as modulators of growth factor signaling in chondrogenesis and gliogenesis. She is also the director of the Joseph P. Kennedy Intellectual and Developmental Disabilities Center at the University of Chicago. She has served as chair of both the Graduate Research and Education Deans and the Postdoc Leaders Group at the Association of American Medical Colleges, and on the board of directors of the Graduate Record Exam and founding advisory board of the National Postdoctoral Association.

Paula Stephan, Ph.D., is professor of economics in the Andrew Young School of Policy Studies at Georgia State University and a research associate of the National Bureau of Economic Research. Her research interests focus on the economics of science. Her empirical work examines the careers of scientists and engineers, the role of the foreign-born in U.S. science, and how the diffusion of information technology affects the productivity of scientists and the process by which knowledge moves across institutional boundaries in the economy. Dr. Stephan currently serves on the National Research Council Board on Higher Education and Workforce. She served a 4-year term on the National

Advisory General Medical Sciences Council of the National Institutes of Health and served on the Advisory Committee of the National Science Foundation's Social, Behavioral, and Economics Program. She was a member of the European Commission High-Level Expert Group that authored the report *Frontier Research: The European Challenge*. Her research has been supported by the Alfred P. Sloan Foundation, the Andrew W. Mellow Foundation, and the National Science Foundation. Dr. Stephan graduated from Grinnell College (Phi Beta Kappa) with a B.A. in economics and earned both her M.A. and Ph.D. in economics from the University of Michigan. She is also author of the book *How Economics Shapes Science* (Harvard, 2012).

Lorraine Tracey, Ph.D., is medical science liaison at Teva Pharmaceuticals. She is a former postdoctoral research associate in the Department of Surgery at St. Jude Children's Research Hospital, where her work focused on the role of NF-kB in treatment response and on rational drug combinations for the treatment of pediatric solid tumors, and also the former director of biological research and development at NanoDetection Technology in Cincinnati, Ohio. Dr. Tracey completed her undergraduate training in human genetics at the University of Dublin, Trinity College, Ireland, and went on to do her Ph.D. at the Spanish National Cancer Center in Madrid, Spain. She has received numerous awards, including the 1999 Bloomer Prize in Human Genetics and the 2003 Spanish Academy of Dermatology and Venereology Prize for Research. She was elected chair of the postdoctoral association council at St. Jude Children's Research Hospital in 2009. In addition, she served on the board of directors of the National Postdoctoral Association from 2010 to 2013, serving as chair from 2012 to 2013.

Michael Turner, Ph.D., is the Bruce V. and Diana M. Rauner Distinguished Service Professor and director of the Kavli Institute for Cosmological Physics at the University of Chicago. Dr. Turner is also president-elect of the American Physical Society (APS). Dr. Turner received his B.S. from California Institute of Technology, his M.S. and Ph.D. from Stanford University, all in physics, and an honorary doctorate from Michigan State University. Dr. Turner helped to pioneer the interdisciplinary field of particle astrophysics and cosmology. His scholarly contributions include predicting cosmic acceleration and coining the term "dark energy," and showing how during cosmic inflation quantum fluctuations evolved into the seed perturbations for galaxies. His honors include the Warner Prize of the American Astronomical Society (AAS), the Lilienfeld Prize of the APS, the Klopsted Award of the American Association of Physics Teachers, the Heineman Prize (with Kolb) of the AAS and American Institute of Physics, and the 2011 Darwin Lecture of the Royal Astronomical Society. Dr. Turner has previously served as chief scientist at Argonne National Laboratory, assistant director for the mathematical and physical sciences at the National Science Foundation, and president of the Aspen Center for Physics. He is a fellow of the APS, American Academy of Arts & Sciences, and the American Association for the Advancement of Science. He is a member of the National Academy of Sciences and has served on its governing board.

Allison Woodall, J.D., is the deputy general counsel of the Labor, Employment and Benefits Group in the Office of the General Counsel for the University of California (UC). Ms. Woodall received her B.A. and her J.D. from the University of California, Berkeley. She previously was a partner at the law firm Hanson, Bridgett LLP before joining UC's Office of the General Counsel in 2010. Ms. Woodall advises on a wide range of legal issues, including collective bargaining, labor-management relations, employment discrimination, personnel issues, and policy interpretation.

Joan Woodard, Ph.D., retired in 2010 from Sandia National Laboratories as executive vice president and deputy director. She served as the chief operating officer from 1999 to 2005. During her 36-year career at Sandia National Laboratories she led the energy technology development programs as well as the national security programs. She holds a Ph.D. in mechanical engineering from the University of California, Berkeley, and a master's in engineering economics from Stanford University. Other directorships include Missouri University of Science & Technology Board of Trustees, Bosque School Board of Trustees, and the New Mexico Women's Forum.